The Open University

BLOCK 5

PLANETARY EVOLUTION

Prepared for the Course Team by

DAVID A. ROTHERY

With contributions from Stephen Blake and Nigel Harris

THE S267 COURSE TEAM

CHAIRMAN

Peter J Smith

C J Hawkesworth

COURSE COORDINATOR

Veronica M E Barnes

COURSE MANAGER

Val Russell

AUTHORS

Andrew Bell

Stephen Blake

Nigel Harris

David A Rothery

Hazel Rymer

EDITORS

David Tillotson

Gerry Bearman

Sue Glover

DESIGNER

Caroline Husher

GRAPHIC ARTIST

Alison George

BBC

David Jackson

The Open University, Walton Hall, Milton Keynes, MK7 6AA.

First published 1993.

Edited, designed and typeset in the United Kingdom by the Open University.

Printed in the United Kingdom by Thanet Press Limited, Margate

ISBN 0 7492 8166 9

This text forms part of an Open University Second Level Course. If you would like a copy of *Studying with the Open University*, please write to the Central Enquiry Service, PO Box 200, The Open University, Walton Hall, Milton Keynes, MK7 6YZ. If you have not already enrolled on the Course and would like to buy this or other Open University material, please write to Open University Educational Enterprises Ltd, 12 Cofferidge Close, Stony Stratford, Milton Keynes, MK11 1BY, United Kingdom.

3.1

S267b5i3.1

S267
HOW THE EARTH WORKS:
The Earth's Interior

BLOCK 5
PLANETARY EVOLUTION

5.1 INTRODUCTION AND STUDY GUIDE

Congratulations on reaching the last Block of S267! We hope that by now you've formed as thorough an impression of 'how the Earth works' as it is possible for us to put across within a half-credit course. In the first Sections of Block 1, you were introduced to some of the evidence bearing on the Earth's formation and the origin of the Solar System in general. Apart from that, we've presented you with a view of the Earth's internal processes as they operate today.

This Block has two principal aims. The first is to assess whether the Earth 'worked' *in the past* in the same way as it does *now*. To address this issue, we raise, without providing detailed evidence, a few important topics concerning the past history of the Earth. In Section 5.2, we look at the plate-tectonic configuration that existed about 200 Ma ago, before most of the present continents had separated and continental crust was instead aggregated into a single 'supercontinent'. This supercontinent had itself been assembled by the collision of a number of earlier continents. The lesson from this is that there were times, even in the relatively recent past (geologically speaking), when the Earth's continental crust was arranged in ways fundamentally different from today, although the general pattern of events appears to have been repeated several times. In Section 5.3, we turn the clock back further to consider how and when the Earth's continental crust was formed, and discover that continental and oceanic crust are products of the recycling of earlier generations of crustal material through the mantle, and that the oldest surviving continental crust has some significant geochemical differences from younger continental crust. These are, to a large extent, a result of higher geothermal gradients in the past, because of greater rates of heat production. In Section 5.4, we look at this heat production, how it affected the thickness of the lithosphere, and its consequences for the nature of mantle-derived melts. In Section 5.5, we peer back to the immediate post-accretion state of the Earth and imagine the effects, on the outer layers of the Earth, of bombardment by meteoritic debris. No direct evidence of this epoch remains on the Earth but the surfaces of planetary bodies that have been geologically less active preserve a more complete record, and we examine how a cratering time-scale has been set up, with the aid of lunar data, that is the basis for estimating the ages of surfaces on other planets.

This leads us to the second aim of the Block, which is to examine whether all comparable planetary bodies function in the same way as the Earth does. In other words, we address the question of *how other planets work*. In Section 5.6, we establish the ground rules for comparisons between the Earth and other bodies in the Solar System, and demonstrate that it is sensible to consider not just the more obviously Earth-like bodies but also all except the smallest of the icy satellites of the outer planets. In subsequent Sections, we consider in turn the Moon (Section 5.7), Venus (Section 5.8), Mars (Section 5.9), Jupiter's most active satellite, Io (Section 5.10), and a few exemplary icy satellites (Section 5.11), concentrating usually on their thermal histories and the nature of their crusts and/or lithospheres, and making comparisons between them and the Earth. Finally, in Section 5.12, we speculate on the lessons we can draw from other planetary bodies in understanding whether or not the Earth's structure and present-day volcanic and tectonic processes are an inevitable consequence of its initial conditions.

We do not promise any definite answers on this last point, nor to the questions raised by the two principal aims of the Block. However, you will probably find it helpful to bear these issues in mind as you read on. Also, you will discover that this Block provides opportunities to revise many of the concepts and skills that were introduced or developed earlier in the Course, so we hope you will find it useful as preparation for the exam.

There are two items of audiovisual material to assist your study of this Block: AV 10, 'Planetary Matters', which accompanies Section 5.5, and VB 07, 'Other Worlds', which you should view alongside Sections 5.6–5.8. We estimate that the total study time for this Block is 2.5 CUEs, divided *approximately* as follows:

Section	CUE
5.1 + 5.2	0.15
5.3	0.15
5.4	0.15
5.5	0.20
5.6	0.25
5.7	0.45
5.8	0.40
5.9	0.20
5.10	0.30
5.11 + 5.12	0.25

5.2   CONTINENTS AND SUPERCONTINENTS

In Block 2, you saw how plate tectonics operates at the present time. The Earth's continental crust is divided into six major land-masses (North America, South America, Antarctica, Australia, Eurasia and Africa), which are almost entirely separated by oceans. Some of these oceans are no more than incipient features, just beginning to grow in the aftermath of continental rifting (e.g. the Red Sea and Gulf of Aden; Block 2, Section 2.5.2), and others (notably the Atlantic) are mature oceans. On the other hand, the Mediterranean is a remnant of a formerly more major ocean basin that has been largely destroyed by continental convergence and collision.

However, the arrangement of the Earth's continents and oceans has not always been in this sort of pattern.

❑   Can you recall an example of a fundamentally different arrangement of continents that you met in Block 2?

■   In Block 2, Section 2.5.4, you were introduced to Gondwanaland, which is the name given to a giant southern continent composed of what subsequently dispersed into South America, Africa, India, Antarctica and Australia.

Figure 2.51 showed how Gondwanaland began to break up soon after 115 Ma ago, and you saw how India moved northwards by some 6 000 km, colliding with the southern edge of Asia about 40 Ma ago. If we look at the generally accepted fit between continents a little further back into the past, we find that at 200 Ma ago *all* the major continents were united. North America and Eurasia (collectively known as **Laurasia**) were joined to the northern edge of Gondwanaland, and the whole assembly formed a single **supercontinent**, which has been given the name **Pangaea** (which means, literally, 'all Earth'). Figure 5.1 shows the stages by which Pangaea broke up.

ITQ 1

What three lines of evidence that you met in Block 2 could have been used to deduce the arrangement in which continents were assembled in Pangaea at 200 Ma ago?

The evidence for the existence of Pangaea is well-nigh incontrovertible. However, as you might expect, the further we attempt to look into the past, the harder it becomes to get a reliable picture of what went on. One reason for this is that virtually all the oceanic crust older than about 160 Ma has been destroyed by subduction (Block 2, Section 2.2.4), and with it the evidence of sea-floor magnetic stripes. However, it is pretty clear that Pangaea had not existed indefinitely. For example, if we look in the region of what is now Western Europe within the continental assembly of Pangaea, geologists can trace belts of metamorphic rock that appear to be the deeply eroded roots of two Himalayan-type mountain chains, produced by continent–continent collision. One, the Variscan belt, which is approximately 340 Ma old, runs along the join between North America and northwest Africa on into southern Europe (Figure 5.2b). The other, the Caledonian belt, which is about 395 Ma old, runs from New England, through Newfoundland and northwest Britain and on along the join between Scandinavia and Greenland (Figure 5.2a). The Urals (a chain of mountains within a continent — Block 2, Section 2.5.4) are now understood to mark the site of the collision that united Europe and the main part of Asia into what we know as Eurasia, at about 300 Ma ago. The obvious conclusion is that Pangaea was assembled by the closure of oceans (at different times) that formerly lay along the lines of these belts. This is supported by a variety of evidence, including apparent polar wander curves (Block 2, Section 2.2), as shown in Figure 5.3, and the occasional survival of slivers of oceanic crust and upper mantle (ophiolites) as thrust slices within collision zones (Block 3, Figure 3.3).

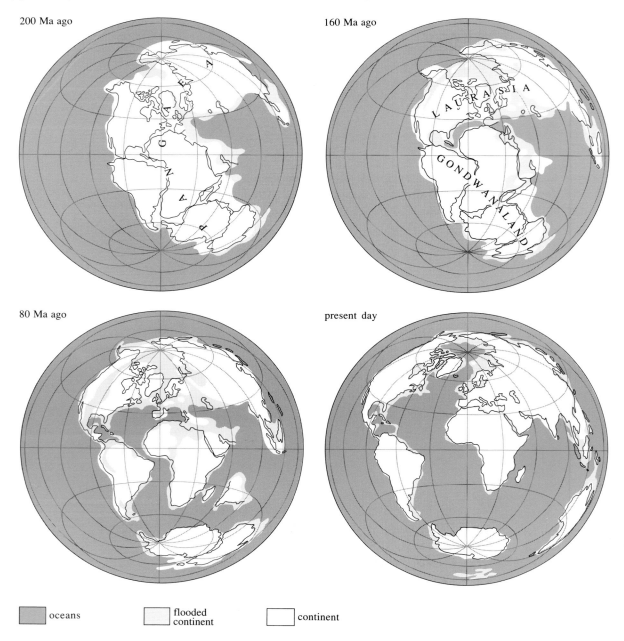

200 Ma ago

160 Ma ago

80 Ma ago

present day

oceans   flooded continent   continent

Figure 5.1   The break-up of Pangaea. At 200 Ma ago, virtually all the Earth's continental crust was united in the single supercontinent of Pangaea. By about 160 Ma ago, this had begun to split apart, with Gondwanaland (in the south) beginning to split from the northern continental mass (North America plus Eurasia, usually known as Laurasia). By 80 Ma ago, Gondwanaland had fragmented (except that Australia and Antarctica remained united) but Laurasia was still intact. Today, Laurasia has broken apart, but India and Africa (formerly parts of Gondwanaland) have collided with the southern rim of Eurasia. Note that this map projection shows the *whole* surface of the Earth.

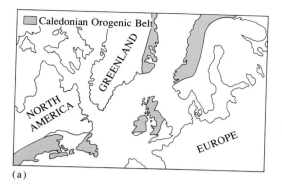

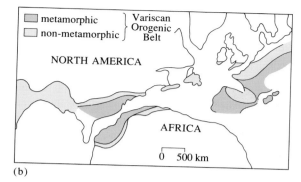

(a)　　　　(b)

ITQ 2

Three important characteristics are revealed in Figure 5.3. The apparent polar wander curves for North America and Europe follow separate tracks until about 390 Ma ago, then they run virtually together until about 150 Ma ago, and they begin to diverge after this time. How would you interpret this?

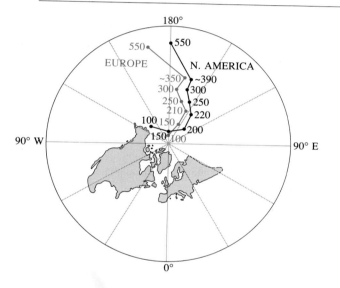

Figure 5.2 The Caledonian (a) and Variscan (b) belts, which are the deeply eroded roots of Himalayan-style mountain belts, marking the sites of ocean closure during the assembly of Pangaea. Land-masses are drawn with no intervening oceans, i.e. as they were before Pangaea (or, at least, Laurasia) broke up.

Figure 5.3 Apparent polar wander curves for Europe and North America fitted together as they lay before the North Atlantic began to open, showing the relative positions of the poles at times (in Ma) in the past.

This is not to say that the North Atlantic opened exactly along the line of the collision where its forerunner ocean had disappeared; fossils and other evidence show that much of Scotland and western Ireland belonged to the American side, and parts of Newfoundland to the European side of the North Atlantic forerunner. To avoid confusion, this North Atlantic forerunner is generally called the **Iapetus Ocean** (pronounced 'Eye-apetus').

An important result emerging from detailed studies is that the closing of the Iapetus Ocean appears to be merely the most recent of a long series of ocean closures to have affected the same region. The geological record shows signs of successive mountain-building episodes (which followed from ocean closure) peaking about every 400–500 million years and traceable as far back as about 2 500 Ma ago, though this becomes increasingly speculative the further back in time we go. In each case, about 100 million years after the peak of the mountain-building episode, rifting and then ocean formation began, so it appears that oceans have existed approximately along the line of the present North Atlantic at several times in the past, the Iapetus Ocean being merely the most recent example. Similar histories can be traced at the margins of other oceans, except for the Pacific. This cycle of continents rifting to form oceans and then the oceans closing to re-unite the continents is known as the **Wilson cycle**, in honour of J. Tuzo Wilson (who first described transform faults: Block 2, Section 2.3.2). A Wilson cycle model is shown schematically in Figure 5.4, which indicates the sequence by which a supercontinent breaks apart and then re-unites. Note that, according to this model, the present Pacific Ocean is the descendant of the hemisphere-wide ocean that surrounded the supercontinent. None of the actual crust forming the Pacific floor has survived from those times, but it has been replaced by younger crust formed by sea-floor spreading at the East Pacific Rise and elsewhere.

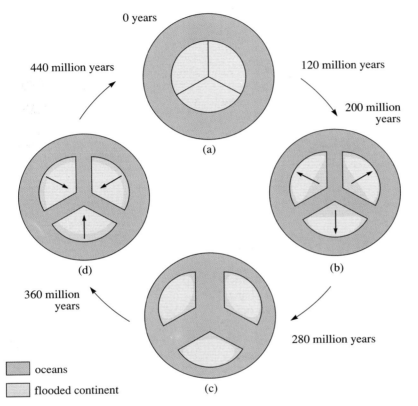

Figure 5.4   The Wilson cycle model of how a supercontinent breaks apart into several separate continents that initially drift apart but then move back together to re-unite the supercontinent about 400–500 million years after it was previously formed. According to this model, the supercontinent of Pangaea was merely the most recent reincarnation of something that had existed several times before. The characteristic areas where the continents tend to be flooded (by shallow seas) can be recognized in the geological record and are consistent with the theory.

The reason why a supercontinent, once formed, does not remain intact is not certain. However, it is thought that essentially it self-destructs, rather than being pulled apart by outside forces. At the sites of previous continent–continent collisions, continental crust is likely to be particularly thick; for example, as you saw in Block 2, Section 2.5.4, a combination of folding and thrusting has led to a crustal thickness in excess of 70 km beneath the Himalayas and parts of Tibet.

❑   A thick layer of continental crust acts as an effective insulator to the mantle-derived heat. Assuming the mantle heat flow to be uniform around the globe, what do you think happens to the heat supplied from the mantle below a thickened area of continental crust?

■   As the heat cannot be conducted upwards through the crust at the same rate as it is supplied, the heat must build up within and below the crust.

This build-up of heat adds to the already high radiogenic heat production within the thickened crust. Two effects now come into play. The first is that, as it becomes hotter, the most-thickened crust expands thermally, becoming less dense, and therefore more buoyant, so that it rises isostatically. This arching of the crust causes stretching across the outside of the arch, which is accommodated by the initiation of normal faults.

The second effect is that the heat trapped below the crust brings the base of the thermal lithosphere closer to the surface (the effect will be exaggerated in continental crust that has been stationary over a hot spot for many tens of millions of years).

The natural outcome of these processes leads to rifting of the kind illustrated in Block 2, Figures 2.42 and 2.44. Note, however, that the splitting apart of continents seems not to occur *exactly* along the join (or 'suture') between them. This is probably because the folding, thrusting and magmatic activity resulting from collision act so as to weld the continents together very intricately, so that the suture is not a line of weakness that the rifting process can exploit. The situation would be complicated still further if any island arcs or exotic terranes (Block 2, Section 2.5.6) had been caught up within the collision zone. There are also instances, such as the 300-Ma-old Urals mountain belt, where continent–continent collision has not (yet) been followed by rifting.

Thus, disregarding exceptions such as the Urals, we now have a basic explanation of how a supercontinent could break up. The African rift system (Figure 2.45) which runs southwards from a triple junction with the Red Sea and the Gulf of Aden may be a reasonable analogue to what happened on a grander scale during the break-up of Pangaea. But what encourages the dispersed fragments of a supercontinent to come back together again?

ITQ 3

You saw in Block 2, Section 2.6.3, that the slab-pull force is probably the most important force influencing the movement of plates. In what way will the buoyancy of the earliest-formed lithosphere of the Atlantic-type oceans shown in Figure 5.4 change by the time stage (c) is reached, and how will this encourage ocean closure?

It would be untrue to pretend that the answer to ITQ 3 provides a complete answer to why supercontinents appear to have been reassembled several times in the manner shown in Figure 5.4. For one thing, it does not explain the apparent longevity of the Pacific Ocean, which, although all crust older than about 160 Ma has been subducted, appears to have been a persistent feature through several cycles.

Oversimplified though it undoubtedly is, the model in Figure 5.4 seems to give a reasonably correct picture of the nature of the Earth's plate tectonics back to around 2 500 Ma ago. If we attempt to peer even further back, into the period known as the **Archaean** (>2 500 Ma ago), the evidence becomes scantier and our view is even more clouded. This appears to have been a time characterized by the assembly of small continental fragments into progressively larger units. Figure 5.5a and b shows one idea of how this may have occurred: small continents underwent extension and volcanism above subduction zones in a process analogous to modern back-arc basin formation by the splitting of island arcs (Block 2, Section 2.3.3). When these small continents collide, the sediments and volcanic rocks of the former extensional basins become compressed into metamorphic complexes known as **greenstone belts** (from the characteristic colour of the metamorphosed volcanic rocks), but the continental 'basement' gneisses suffer much less deformation. Both the greenstone belts and the continental basement are intruded by granites during and after collision. This story may sound convincing, but if you consider the geological evidence on which it is based (e.g. Figure 5.5c) you will realize that the details of crustal assembly in the Archaean are far from being well-documented. In the next Section, we will turn to the related problem of the age of the Earth's continental crust, and how and when it grew.

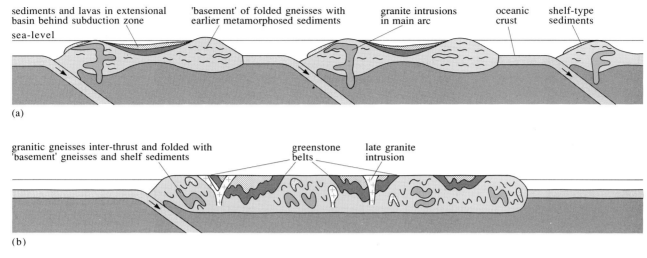

(a)

(b)

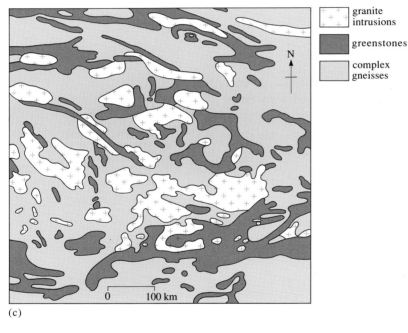

(c)

	granite intrusions
	greenstones
	complex gneisses

Figure 5.5 (a, b) Plate tectonic model for the growth of continental crust in the Archaean. (c) Typical pattern of rock types within a tract of continental crust that has remained stable since the Archaean (in this example, from the Superior craton, Canada, which is drawn at a larger scale than parts (a) and (b)). The gneisses are understood to represent the pre-existing continental crust in Figure 5.5a; the greenstones are the remains of the sediments and lavas formed in the extensional basins; and most of the granites are intrusions that post-date the collisions.

SUMMARY OF SECTION 5.2

The supercontinent of Pangaea included virtually all the present continental crust. It was assembled between about 390 and 300 Ma ago, and had begun to split apart into Gondwanaland and Laurasia by about 160 Ma. These two major continental masses have been breaking apart into smaller fragments ever since. Before Pangaea came together, the Iapetus Ocean separated what are now North America and Europe along the approximate (but not exact) line of the present North Atlantic.

Repeated opening and closing of Atlantic-type oceans and the associated formation and break-up of a supercontinent is known as the Wilson cycle, and can be traced (with increasing uncertainty through time) back to about 2 500 Ma ago. The break-up of a supercontinent can be attributed to thermal doming and rifting of crust that had previously been thickened by continent–continent collision. One factor that probably contributes to the reassembly of a supercontinent is the slab-pull force following initiation of subduction in Atlantic-type oceans when the oceanic lithosphere has become old, cold and dense enough.

Prior to 2 500 Ma ago, the picture appears to be one of assembly of continental crust from smaller pieces.

OBJECTIVES FOR SECTION 5.2

When you have completed this Section, you should be able to:

5.1 Recognize and use definitions and applications of each of the terms printed in the text in bold.

5.2 Understand the nature of the evidence for, and some of the theories to explain, the nature of plate-tectonic activity and continental reconstructions in the past.

5.3 Be able to relate processes described in this Section to those you have met earlier in the Course, particularly in Block 2.

Apart from Objective 5.1, to which they all relate, the three ITQs in this Section test the Objectives as follows: ITQ 1, Objective 5.2; ITQ 2, Objectives 5.2 and 5.3; ITQ 3, Objective 5.3.

You should now do the following SAQs, which test other aspects of the Objectives.

SAQS FOR SECTION 5.2

SAQ 1 *(Objectives 5.1 and 5.2)*

Bearing in mind the age of the oldest surviving oceanic crust, can you explain why it is not possible to track the movements of continental fragments prior to the assembly of Pangaea as easily as this can be done subsequent to the break-up of Pangaea?

SAQ 2 *(Objective 5.3)*

By comparison with Block 2, Figures 2.42 and 2.44, can you suggest why there should be widespread flooding of the particular continental regions indicated at stage (b) in Figure 5.4?

5.3 GROWTH OF CONTINENTAL CRUST

Earth scientists have accumulated a vast database on the composition and properties of the continental crust, partly because of its obvious importance in providing the environment for the evolution of terrestrial life and partly because, of all the Earth's layers, the continental crust is the most accessible. It may therefore surprise you that there remains a considerable debate about when the continental crust first formed and how rapidly it evolved through the 4 500 million years of Earth history.

The age of continental crust is extremely variable and the search for the Earth's earliest crust has gone on almost since the science of geochronology began. During the 1960s, Greenland was often hailed as the site of the oldest known crust, providing Rb–Sr isochron ages of 3 700 Ma (Figure 4.31, Block 4). The fact that today isotopic ages from crustal material stretch back to 4 200 Ma tells us less about the age of the crust than about the increasingly sophisticated measuring tools that geochronologists use. Geochronological ages older than about 3 800 Ma are obtained from small regions within zircon crystals a few micrometres across, using an extremely sophisticated mass spectrometer called an ion microprobe. These old ages certainly provide evidence that zircons grew early in the Earth's history, but zircons as old as 4 500 Ma are found in the mantle and also in meteorites. Therefore, ancient zircons may be entrained in a solid state into younger magma, and all we can say with confidence about the rock in which they sit is that it must have been formed after the time at which the small part of the analysed zircon formed.

The exact date of the Earth's earliest crust depends partly on our definition of crust. If we take as our starting point a primitive Earth enveloped by a global magma ocean (Block 1, Section 1.3.3), the first solid surface on the Earth was probably no more than a chilled skin on the surface of the magma ocean (just as a volcanic lava lake rapidly forms a solid crust). This **chilled crust** would have been compositionally the same as the underlying magma, and so was not a true crust in the geochemical sense of being differentiated from the mantle. Furthermore, it would have been frequently punctured by impacts and torn apart from below by vigorous convection. However, as the mantle began to cool and crystallize, differentiation would result in the least dense (and earliest crystallized) minerals rising towards the surface. This process would have formed the Earth's **primary crust**, so-called because it was the first differentiated crust to grow following accretion. The primary crust was probably in existence within a few million years of the Earth's formation, but has not survived in any recognizable form. In part, this can be attributed to its destruction by the same (but waning) forces that would have made the chilled crust unstable, but in addition it has been recycled and destroyed by processes leading to the formation of younger types of crust, known as secondary and tertiary crust. **Secondary crust** is formed by partial melting of the mantle.

❑   Is secondary crust being formed on the Earth today, and if so what is it?

■   Partial melting of the mantle is the process by which the Earth's oceanic crust is generated, as you should realize from Block 3, Section 3.5.

Oceanic crust is still being generated today, and is also being destroyed by subduction. Hence, there is no oceanic crust surviving beneath the oceans older than about 160 Ma, although a few small and highly deformed slivers of older material have survived on land in the form of ophiolites (Block 2, Section 2.4.2). If we want to sample the oldest surviving crust on Earth, we must look for the oldest **tertiary crust**. This is the continental crust, which, as described in Block 4, is formed from a mixture of pre-existing crust and mantle material by a combination of partial melting and fractional crystallization processes. You will see later in this Block that these definitions of different types of crust are applicable equally well to other planetary bodies. For some of these, primary crust has survived, whereas others are covered by secondary or tertiary crust.

- ■ Turning our attention back to the Earth for the time being, why would tertiary crust not be destroyed eventually?

- ❑ Continental crust has lower density than oceanic crust and therefore cannot be subducted at convergent margins (Block 4, Section 4.2).

Tertiary crust, then, may be preserved through geological time and it is therefore reasonable to ask the question: 'when did the Earth's continental crust form?' Did it all form in the first 1 000 million years of the Earth's existence, or has it formed continually through geological time? Well, we know that some continental crust has formed within the past 100 Ma in the Andes (Block 4, Section 4.3.8), and we can therefore infer that crustal growth did not stop dead after the first few hundred million years of Earth history. We reached our conclusions about crustal growth in the Andes from comparing the initial Sr-isotope ratios of igneous rocks with those of the evolution curves of Sr-isotope ratios for the upper mantle and for potential source regions in the crust. This technique can be extended to estimate the growth periods of the continental crust world-wide. Sr-isotopes are not ideal for this approach because the change in Rb/Sr ratios that occurs when continental crust is formed can also occur in the mantle independently of crust-forming events. For this reason, other isotopic systems are harnessed (such as ^{147}Sm $\rightarrow$ ^{143}Nd) that are based on the fact that element fractionation (in this case Sm/Nd) occurs *only* when crust is formed. Although the present-day global growth rate of continental crust has been estimated to be about 1 km^3 a^{-1}, there are enormous uncertainties in such estimates. This is partly because only a small proportion of igneous rocks have been analysed isotopically, and partly because we have no precise information on how the crust varies in age both below and above the present level of exposure. The majority of samples available for analysis are taken from this rather arbitrary level in the crust.

Four models for the growth of continental crust are plotted in Figure 5.6. Cursory inspection will show that these curves differ in many respects, reflecting the contrasting models on which they are based. The details of these models lie beyond the scope of this Course, but the Figure does illustrate how imprecise understanding of crustal growth is

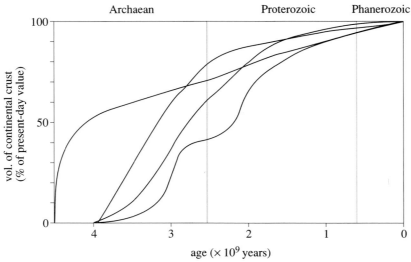

Figure 5.6 Four models (published since 1981) for the growth of continental crust through geological time.

at the time of writing. Even so, three conclusions appear to be broadly accepted:

1   The earliest continental crust to be preserved was formed 3 700–4 500 Ma ago.

2   At least 50% of the Earth's continental crust had formed before 2 000 Ma ago.

3   The Earth's continental crust continues to grow, albeit at a slower rate than in the past.

Earth history can be divided into three periods: the Archaean, which covers the early history until 2 500 Ma ago; the Proterozoic, from 2 500 to 570 Ma; and the Phanerozoic, which covers the most recent period during which life evolved (the Proterozoic/Phanerozoic boundary was originally defined as the time at which life began, but fossils have now been traced back much further in time). The evolution curves of Figure 5.6 tell us that by the end of the Archaean 40–80% of the present-day crust had already formed. Is there evidence that the processes that formed this early crust were in any way different from the processes that form crust today at destructive plate margins? This is an important question because the rate at which continental crust is formed is much slower today than during the Archaean, so the mechanisms of crustal growth which we infer for present-day margins may not be very significant for planetary evolution over the broad sweep of its history.

One way of answering this question is to compare the geochemistry of magmas formed in the Archaean with those formed more recently. In 1990, a geochemical study of magmas of Archaean age noted some critical differences in trace-element geochemistry with magmas formed at island arcs in the past 65 Ma (Figure 5.7).

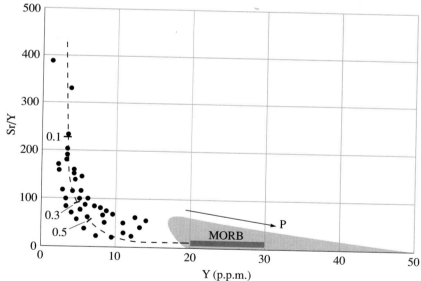

Figure 5.7 Sr/Y versus Y plot for plutons of intermediate composition formed during the Archaean (●) and for volcanics from island arcs < 65 Ma old (shaded). Dashed line indicates partial melting of eclogite (ticked at F = 0.1, 0.3, 0.5). Arrow indicates trend of liquid compositions resulting from fractional crystallization of plagioclase.

The great majority of recent island-arc volcanics lie in a field with low Sr/Y ratios and high Y concentrations (> 15 ppm). Since Y behaves compatibly if garnet is present (K_D = 11 between garnet and melts of intermediate composition), the high Y in the island-arc magmas suggests that there was no garnet present in the source from which they have been derived. We know from Block 4 that fractional crystallization of plagioclase is an important part of the process for deriving andesitic and dacitic magmas from basalts at destructive margins. For intermediate magmas, Sr is compatible with plagioclase (K_D = 4.4) and Y is incompatible with plagioclase (K_D = 0.06) and hence its removal from a magma will drive the magmatic composition to higher Y concentrations and lower Sr/Y ratios (P on Figure 5.7).

In contrast, Archaean plutonic rocks of intermediate composition (Figure 5.7) have much lower Y concentrations (< 15 ppm) and the lowest Y samples have high Sr/Y ratios (> 100). The first conclusion we can draw from this is that garnet was present in the source of these magmas, reducing the Y concentrations of the melts. The second conclusion is that the magmas appear to be unaffected by fractional crystallization of plagioclase. The elevations of Sr/Y ratios in Archaean compositions are not just due to a decrease in Y, but to an increase in Sr as well, which would be inconsistent with removal of plagioclase from the melt.

These geochemical trends cannot be modelled by fractional crystallization of basalts generated by partial melting of the mantle. Rather, they require a garnet-rich and plagioclase-free source such as eclogite. Eclogite, as you may recall from Blocks 3 and 4 and Kit rock sample 4, is made up of roughly equal proportions of garnet and clinopyroxene. For such an assemblage, Sr is incompatible ($D \sim 0.2$) and Y compatible ($D \sim 10$).

ITQ 4

Given that for a typical eclogite Sr = 260 and Y = 40 ppm, deduce the Y and Sr/Y values that result from a 10% melt proportion ($F = 0.1$) using the equation for partial melting (Equation 3.13, Block 3; or Equation 4.3, Block 4).

ITQ 5

What conditions favour melting of the subducted slab at present-day destructive margins?

SUMMARY OF SECTION 5.3

The earliest continental crust to be preserved is termed 'tertiary crust' and its age lies between 3 700 and 4 500 Ma. By at least 2 000 Ma ago, 50% of the Earth's continental crust had formed. During the Archaean, magmas resulted predominantly from partial melting of eclogites and had low Y concentrations and high Sr/Y ratios compared with magmas formed at Phanerozoic island arcs.

OBJECTIVES FOR SECTION 5.3

When you have completed this Section, you should be able to:

5.1   Recognize and use definitions and applications of each of the terms printed in the text in bold.

5.4   Interpret crustal growth curves for the evolution of continental crust.

5.5   Understand the contrasting trace-element characteristics of Archaean and Phanerozoic magmas and relate these to their contrasting source regions and thermal regimes.

Apart from Objective 5.1, to which they all relate, the two ITQs in this Section test Objective 5.5.

You should now do the following SAQ, which tests other aspects of the Objectives.

SAQ FOR SECTION 5.3

SAQ 3

Is it true that the Earth's continental crust has grown at a steadily diminishing rate since the formation of the planet *c*. 4 500 Ma ago?

5.4   THE THERMAL HISTORY OF THE EARTH'S INTERIOR

In Blocks 1 to 4, we saw that the transfer of heat energy within the Earth drives plate tectonics, mantle plumes (and hot spots) and magmatism — the surface manifestations of a dynamic planetary interior. In Section 5.3, we used geochemical evidence to conclude that the geothermal gradient may have been higher in the Archaean than it is at present. This would mean that there was either a thinner lithosphere (or at least a thinner conducting layer — see Block 1, Section 1.13) or a hotter interior (or both) in the past. Testing the former idea is not simple because there is no obvious way of measuring the thickness of the lithosphere in the distant geological past. However, it is possible to assess how the Earth's internal temperature may have changed. Heat is lost from the Earth's interior by convection, magmatism and conduction, but heat is also being gained by radioactive decay of certain isotopes (^{40}K, ^{235}U, ^{238}U, ^{232}Th). So, is the Earth's interior heating up or is it cooling down?

❑   How can we answer our question about the heating, or cooling, rate of the Earth?

■   Basically, there are two approaches. The first would be to look for evidence in the geological record for signs of the Earth's interior having had a different temperature. For example, we would expect the amount of basaltic magmatism to have been greater in the past if the mantle's temperature was higher (Block 3, Section 3.5.1). Secondly, we could try to quantify the Earth's heat budget to see if the amount of heat escaping exceeds that being generated internally, implying an overall cooling. Ideally, both approaches should be followed and their results compared. This is what we'll do here.

5.4.1   THE EARTH'S PRESENT HEAT BUDGET

The Earth loses heat and generates heat, the overall effect being that:

net rate of heat gain = rate of heat addition − rate of heat loss

(Equation 5.1)

If the net rate is negative, then the Earth is cooling. If the net rate is positive, then it is heating up. Put another way, if the ratio of the internal heat generation rate to the rate of the surface heat loss is less than 1, the Earth is cooling; if the ratio is greater than 1, it is heating up. This ratio is know as the **Urey ratio (Ur)** in recognition of the eminent American scientist Harold C. Urey:

$$Ur = \frac{\text{rate of internal heat generation}}{\text{rate of surface heat loss}}$$

(Equation 5.2)

ITQ 6

In Block 1, Section 1.12, we noted that the Earth's surface heat flow amounts to about $4 \times 10^{13}\,W$ and that one estimate of the total rate of heat generation is $2.2 \times 10^{13}\,W$ (Block 1, ITQ 48(b)). On this basis, what is the Urey ratio of the Earth at the present day?

Other estimates of the Urey ratio, based on different assumptions about the abundance of heat-producing elements in the Earth, fall in the range 0.5 to 0.8 (Block 1, Section 1.12.1). These results lead to the same conclusion that comes from your estimate of Ur in ITQ 6 — the Urey ratio is less than 1, so the Earth must be cooling down.

A simple calculation allows us to estimate just how fast the Earth's interior temperature is falling. However, because the Earth is layered, its interior is not kept at a well-mixed, homogeneous, temperature; the inner core is hotter than the outer core, which is hotter than the mantle. As the temperature of the mantle has a bearing on the production of basaltic lavas that are incorporated into the geological record (which we can interpret in terms of the mantle's temperature, as in Block 3), it is useful to concentrate on the mantle's heat budget. Again we can write:

net rate of heat gain of mantle = rate of heat addition to mantle − rate of heat loss from mantle   (Equation 5.3)

❑   What are the sources of heat addition to the mantle?

■   There are two sources — internally generated radiogenic heat and heat transported by conduction from the core across the core mantle boundary into the base of the mantle.

Heat from the core drives the convective plumes associated with hot-spot volcanoes and these are estimated to carry about $2 \times 10^{12}\,\mathrm{W}$ into the mantle. This is smaller by a factor of 5 than our estimate of the internal heat production in the mantle ($10^{13}\,\mathrm{W}$, Block 1, ITQ 48).

❑   What are the means of heat loss from the mantle?

■   There are three mechanisms. Plate creation, at constructive margins, involves the extraction of hot magma from the mantle whilst plate destruction at subduction zones involves cold lithosphere sinking into the mantle, cooling it down. Thus, plate tectonics acts to cool the Earth's interior. The second mechanism of heat loss occurs away from plate margins. Here, heat is conducted through the lithosphere. Thirdly, mantle plumes deliver heat to the base of the lithosphere to be lost by conduction and hot-spot magmatism.

The rate at which heat is being removed from the mantle is estimated as $3.6 \times 10^{13}\,\mathrm{W}$, with most being lost due to plate recycling, and hot spots making the smallest contribution (Figure 5.8).

❑   Is the mantle cooling down?

■   Yes. The total heat loss ($3.6 \times 10^{13}\,\mathrm{W}$) is greater than the total heat input ($10^{13} + 0.2 \times 10^{13} = 1.2 \times 10^{13}\,\mathrm{W}$), resulting in a heat loss from the mantle of $2.4 \times 10^{13}\,\mathrm{W}$.

To express this heat loss in terms of a drop in temperature, we need to make use of the fact that:

$$\Delta E = mc\Delta T \qquad \text{(Equation 5.4a)}$$

i.e.

change in heat energy, ΔE (J) = mass of cooling material, m (kg) × specific heat capacity, c ($\mathrm{J\,kg^{-1}\,K^{-1}}$) × change in temperature, ΔT (K)
(Equation 5.4b)

The rate of change in heat energy can be expressed as follows:

Q ($\mathrm{J\,s^{-1}}$, or W) = mass of cooling material, m (kg) × specific heat capacity, c ($\mathrm{J\,kg^{-1}\,K^{-1}}$) × rate of change in temperature, T' ($\mathrm{K\,s^{-1}}$)
(Equation 5.5a)

i.e.

$$Q = mcT' \qquad \text{(Equation 5.5b)}$$

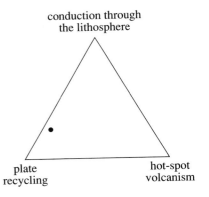

Figure 5.8 The plotted point on this schematic ternary diagram indicates the approximate relative contributions made by plate recycling, conduction and hot spots to the Earth's heat loss.

Note that in obtaining Equation 5.5 from Equation 5.4, we have made the assumption (which turns out to be reasonable) that the proportional change in the mass of the mantle due to extraction of magma is insignificant in comparison with its proportional change in temperature.

ITQ 7

Given that the mantle has a mass of 4.06×10^{24} kg and a specific heat capacity estimated as $1\,000\,\mathrm{J\,kg^{-1}\,K^{-1}}$, calculate the current rate of temperature change in the mantle. Express your result in degrees per 10^9 years (i.e. gigayears, Ga) given that 1 year contains 3.15×10^7 s.

Your answer to ITQ 7 is indicative rather than absolutely accurate because, as you should be aware, the estimated heat budget and specific heat capacity are open to debate. Alternative values give cooling rates between 50 and 200 K Ga^{-1} but the conclusion is always the same: the mantle is cooling very very slowly (you might like to see just how slowly by calculating the decrease in temperature that has occurred over your lifetime).

5.4.2 THE EARTH'S HEAT BUDGET AND TEMPERATURE IN THE GEOLOGICAL PAST

Having estimated the current cooling rate of the mantle (50–200 K Ga^{-1}), it seems at first glance a simple matter to find the mantle's temperature at any time in the past. But let's stop to think for a moment.

In the past, the amount of heat-producing isotopes must have been greater than at present, so the current rate of heat generation is less than in the past (Figure 5.9). We should, therefore, expect an even hotter mantle than that anticipated from the present cooling rate (assuming the same rate of heat loss).

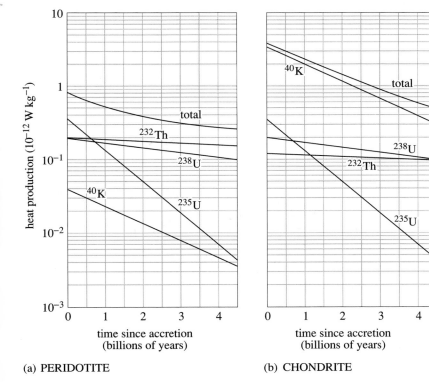

(a) PERIDOTITE   (b) CHONDRITE

Figure 5.9 Variation in the rate of heat production through time for the long-lived isotopes ^{40}K, ^{235}U, ^{238}U and ^{232}Th assuming the composition of (a) peridotite (Block 1, Table 1.12) and (b) chondritic meteorites (Block 1, ITQ 48(a)).

On the other hand, the rate of heat loss is strongly linked to the vigour of mantle convection. A hot mantle will have a low viscosity, and that means vigorous convection (remember that viscosity appears in the bottom line of the Rayleigh number — Block 1, Section 1.14.2) and hence more rapid heat loss in the past.

The rates of heat generation and of heat loss must have been greater in the geological past, but mathematical models of the evolving heat budget suggest that heat loss has always outpaced heat generation (Ur always less than 1), such that the mantle has always been cooling down. Such models (e.g. Figure 5.10) indicate a relatively rapid cooling over the first billion years or so, tailing off as the increasing viscosity of the mantle causes a decrease in the convective cooling rate.

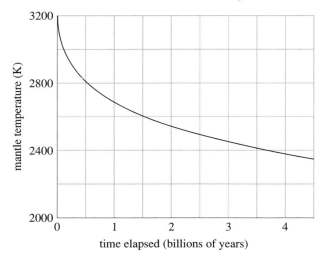

Figure 5.10   An example of the results from a mathematical model of the thermal evolution of the mantle.

5.4.3   GEOLOGICAL EVIDENCE FOR THE EARTH'S THERMAL HISTORY

In the previous two Sections (5.4.1 and 5.4.2), an assessment of the Earth's heat budget from a physical standpoint led us to suspect that the Earth is slowly cooling. This conclusion was based on a number of uncertain estimates of heat production and specific heat capacity, and the simplification that the entire mantle is well-stirred rather than stratified into upper and lower regions which may behave independently. This means that our physical model is speculative. In this Section, we shall make an independent estimate of the Earth's thermal history using geological information, and then compare our conclusions with those from the physical model.

❑   What might we expect the geological consequences of a hotter Earth to be?

■   A hotter interior would lead to higher surface heat fluxes and thinner lithosphere, perhaps leading to a different style of plate tectonics, possibly involving a large number of small, fast-moving plates. Another possibility is that subducted lithosphere would have been hotter than at present, so that magmatism at destructive margins could have been dominated by the products of slab melting rather than of wedge melting. We found chemical evidence for this scenario in Section 5.3. Lastly, a hotter mantle should cause partial melting beneath ocean ridges and hot spots to occur at deeper levels, and yield larger melt fractions (Block 3, Section 3.5) than at present.

So, can we find evidence of this in the geological record?

Unfortunately, the oceanic crust is everywhere younger than 0.2 Ga and its chemical composition is indistinguishable from present day MORB, so that any change in the temperature of the upper mantle over the past 0.2 Ga cannot be resolved.

Turning the geological clock back almost as far as possible, we encounter lavas originally erupted about 3.5 Ga ago (according to radiometric dating) but now exposed in the Komati river valley within the Barberton Mountain Land of South Africa. These rocks are interesting because they include flows which have much higher MgO contents than lavas of the present day (Table 5.1). Formally, these rocks are known as **komatiites** and are defined as ultrabasic magmas with more than 18% MgO.

Table 5.1 Chemical compositions of three komatiites (1: from Barberton Mountain Land, S. Africa; 2: from Munro Township, Ontario, Canada; 3: from Gorgona Island, Colombia); and a primitive MORB (column 4) in mass %. Age (in Ga) is given in brackets.

	1 (3.5)	2 (2.7)	3 (0.15)	4 (0)
SiO_2	44.86	40.8	45.3	48.8
TiO_2	0.28	0.25	0.60	1.15
Al_2O_3	3.12	10.0	10.6	15.9
FeO	11.53	9.14	10.9	9.8
MnO	0.18	0.16	0.18	0.17
MgO	33.01	23.3	21.9	9.7
CaO	3.76	6.86	9.25	11.2
Na_2O	0.01	0.23	1.04	3.4
K_2O	0.01	0.07	0.02	0.08
P_2O_5	0.03	0.02	–	–

Komatiites are typically (but not exclusively) extremely old rocks, being most common in Archaean (i.e. older than 2.5 Ga) greenstone belts. It is therefore interesting to consider what type of conditions would have favoured the production of high-MgO liquids in the early periods of Earth history.

The MgO content of a basalt is significant because it reflects not only the magma's chemical composition but also its liquidus temperature (Block 3, Section 3.6.2). Melting experiments have shown that this also applies to lavas that are richer in MgO than basalts, such as komatiites (Figure 5.11). This means that we can relate a change in chemical composition over time to a change in temperature over time.

ITQ 8

Estimate the liquidus temperature of the lavas listed in Table 5.1.

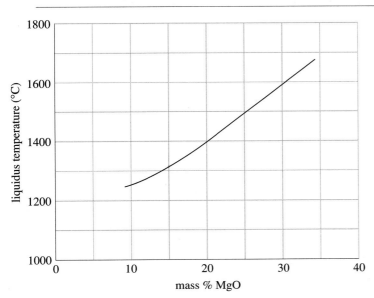

Figure 5.11 Liquidus temperature at atmospheric pressure ($10^5 \, Nm^{-2}$) of basic and ultrabasic magmas as a function of MgO content.

Your answer to ITQ 8 should show that Archaean komatiites, with liquidus temperatures of up to 1 660 °C, were much hotter than primitive basalts from current mid-ocean ridges. Now, because the thermal regime of the mantle determines the chemical and thermal characteristics of magmas produced by partial melting of the mantle (Block 3, Section 3.5), we can state that hot primary (or near-primary) magmas must indicate a hot mantle source region. Thus, Figure 5.12 shows the adiabatic path followed by mantle as it decompresses beneath a mid-ocean ridge and produces parental MORB magmas with liquidus temperatures close to 1 250 °C (cf. Block 3, Figure 3.32b). The analogous path which produces komatiite with a one atmosphere liquidus temperature of 1 650 °C is also shown in Figure 5.12. Although this path is complicated by the pronounced pressure-related shape of the peridotite liquidus and solidus at high pressure, the basic conclusion is amply clear — komatiites require a mantle which is much hotter than that implied by modern magmas.

Having got this far in interpreting the significance of komatiites, it is frustrating to find that the environment in which komatiites were erupted is equivocal. Some geologists believe they were associated with hot spots, whilst others view them as integral parts of oceanic-type crust produced at divergent plate margins. The reason for the uncertainty is that these ancient rocks have been deformed by later phases of geological upheaval, scrambling the geological clues to their original affinity. The most modern recognized komatiites (Gorgona, Table 5.1, column 3) are, however, plume-related. These high-temperature magmas probably erupted when the plume responsible for the basaltic magmas forming the Caribbean Plateau (Block 3, Figure 3.39) first reached the top of the mantle. Plumes are also the likely explanation of Archaean komatiites because the mantle temperatures that they imply (Figure 5.12) are so high that the Archaean mantle would have experienced much more melting and melt extraction than seems compatible with chemical evidence for the mantle's compositional evolution.

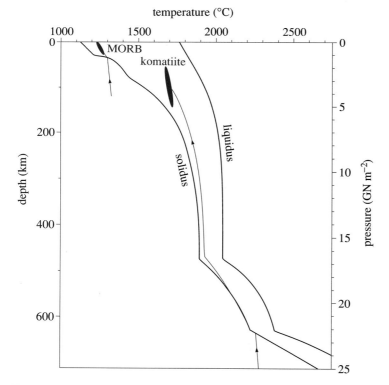

Figure 5.12   Phase diagram of peridotite, showing its solidus and liquidus, with adiabatic rise paths of the mantle leading to generation of MORB and komatiite. Toned areas show approximate pressure and temperature conditions of melting and segregation associated with MORB and komatiite. (1.0 GN m^{-2} = 1.0 × 10^9 N m^{-2}.)

In conclusion, we can state that the decrease in the MgO content of the most primitive magmas through the history of the Earth implies a cooling of the mantle and/or mantle plumes during this time.

SUMMARY OF SECTION 5.4

The thermal evolution of the Earth can be constrained by its present heat budget and by changes in MgO contents (and hence liquidus temperatures) of primary magmas. Estimates of the current Urey ratio (Ur = rate of internal heat generation/rate of surface heat flow) are less than 1, implying an overall loss of heat energy from the Earth. The mantle's estimated heat budget and thermal properties suggest that it is presently cooling at about 50 to $200\,\text{K}\,\text{Ga}^{-1}$, but the rate was probably greater in the past owing to a hotter and hence more mobile mantle allowing convection to outpace high radiogenic heat production. Independent evidence for a cooling Earth comes from the observation that the Earth's hottest magmas (komatiites) are typically found in rock sequences older than 2.5 Ga (i.e. Archaean).

OBJECTIVES FOR SECTION 5.4

When you have completed this Section, you should be able to:

5.1 Recognize and use definitions and applications of each of the terms printed in the text in bold.

5.6 Use simple heat balance models to describe and discuss the heat budget of the Earth (or any other planet).

5.7 Use information on the MgO contents of primary magmas to constrain the thermal conditions of magma genesis.

Apart from Objective 5.1, to which they all relate, the three ITQs in this Section test the Objectives as follows: ITQs 5 and 6, Objective 5.6; ITQ 7, Objective 5.7.

You should now do the following SAQs, which test other aspects of the Objectives.

SAQS FOR SECTION 5.4

SAQ 4 *(Objective 5.6)*

Assume that the mantle (mass $4.06 \times 10^{24}\,\text{kg}$) has an internal heat generation rate identical to that of chondritic meteorites (0.5×10^{-8} $\text{mW}\,\text{kg}^{-1}$; Block 1, ITQ 48(a)), the heat input from the core is $2 \times 10^{12}\,\text{W}$ and heat flux out of the mantle is $3.6 \times 10^{13}\,\text{W}$.

(a) Explain whether the mantle with these properties will be cooling down or heating up.

(b) Calculate the rate of temperature change in the mantle (assume the specific heat capacity is $1\,000\,\text{J}\,\text{kg}^{-1}\,\text{K}^{-1}$).

(c) Explain why your answer to part (b) is different from the cooling rate obtained in Section 5.4.1 (ITQ 6).

SAQ 5 *(Objective 5.7)*

Archaean komatiites occur interbedded with basalt lavas. Use the knowledge gained from Blocks 3 and 4 to suggest three explanations for the origin of these basalts.

5.5   THE DEEPS OF TIME

You saw in the previous Section that during the Archaean the Earth's surface heat flow was probably significantly greater than at present, and this and earlier Sections looked briefly at the possible nature of changes in crustal volume, lithospheric thickness and the style of plate-tectonic activity over time. In this Section, we shall attempt to peer even further into the past to try to discern what conditions were like on the Earth's earliest solid surface. In doing so, you will discover that the best evidence for conditions so long ago comes not from the Earth itself (where crustal recycling, erosion by wind and water, burial by sediments, and a prolonged history of volcanic and tectonic activity have conspired to obliterate all but the most cryptic traces) but from our nearest neighbour, the Moon.

If you cast your mind back to the description of the primitive Earth in Block 1, Section 1.3.3, and compare it with the more evolved Earth considered in Sections 5.3 and 5.4 of this Block, you will see that there remains a significant time gap to be filled. In Block 1, we left the primitive Earth as a hot body probably covered by a magma ocean resulting from accretional heating, and inevitably subject to a continued (but presumably dwindling) rain of planetesimals or other debris that had survived the main episode of planetary accretion. The data presented in Sections 5.3 and 5.4 relate to an Earth that had already lost its primary crust and had been resurfaced by partial melts from the mantle (giving rise to oceanic crust), and recycling of oceanic crust was leading to the growth of continental crust. The direct evidence of conditions on the Earth in the interim (a period of probably several hundred million years) has been destroyed; remember that even though zircon grains older than 3 800 Ma have been identified, this does not mean that the rocks in which they occur are just as old. We would be obliged to rely on theoretical models to describe what went on in the deep past, except for our good fortune in having a large satellite in orbit around the Earth, on whose surface is preserved abundant evidence of the external influences that must have affected the early development of the Moon and the Earth in equal measure. In studying the Moon's early history, our quest to understand surface conditions on the young Earth becomes an inquiry into how the intense bombardment of a planetary surface in the aftermath of accretion subsided to the present, evidently much quieter, situation.

As you will see in Section 5.7, over the past 4 billion years the Moon has been much less active geologically than the Earth (essentially because its smaller size, and hence greater surface-to-volume ratio, allows it to lose heat more rapidly than the Earth, so it has developed a much thicker lithosphere). As a result of the Moon's inactivity, what is widely believed to be the Moon's earliest differentiated crust has survived in the form of the **lunar highlands** (Figure 5.13). These occupy about 60% of the near-side of the Moon and almost all the lunar far-side. We know from direct sampling that the lunar highlands are similar to a terrestrial igneous rock-type known as **anorthosite**, so-called because it is rich in anorthite crystals. The primitive Moon is likely to have been covered by a magma ocean, for the same reasons as the primitive Earth (Block 1, Section 1.3.3), and anorthite crystals are generally interpreted to have been the first phase to have crystallized from the lunar magma ocean. Being less dense than the melt, these crystals would have tended to rise, leading to the growth of the Moon's first differentiated crust.

❑   According to the terminology established in Section 5.3, how would you categorize this crust?

■   As the lunar highlands are the first differentiated crust, this must be the primary crust of the Moon.

Figure 5.13   An oblique view across part of the Moon, as seen from *Apollo 17*. Lunar highlands terrain occupies the left of the view, and younger lavas cover the lower-lying region to the right (which is one of the lunar maria, discussed later in the text).

The oldest anorthosite samples returned to Earth from the lunar highlands have been dated radiometrically at about 4 420 Ma old, and so are somewhat older than the earliest-preserved crustal rocks on Earth (Section 5.3). The lunar highlands show very few signs of tectonic deformation, but are covered densely by craters, almost all of which are clearly **impact craters** produced by the impact into the Moon's surface of projectiles such as large meteorites, comets, asteroids and the like, which are expected to strike the Moon at speeds of about 10–40 km s^{-1}. The processes by which a crater is excavated as a result of an impact (first of all, compression due to the initial shock wave, and then excavation as material is flung out) are now well understood, and are illustrated in Figures 5.14 and 5.15. You will see film of one of the experiments that enabled scientists to reach this understanding at the beginning of VB 07, 'Other Worlds'. If convenient, you might like to view the first 9$^1/_2$

minutes of the video now, but there will be no harm done if you prefer to save it all until you have reached Sections 5.6–5.8. Points worth noting here are:

- impact craters are circular, except in rare cases when the projectile strikes the surface extremely obliquely;

- the ultimate diameter of an impact crater is 1–2 orders of magnitude greater than that of the projectile that causes it;

- lunar craters 1 km in diameter probably took about 1 s to form, whereas 100-km-diameter craters probably took about 100 s.

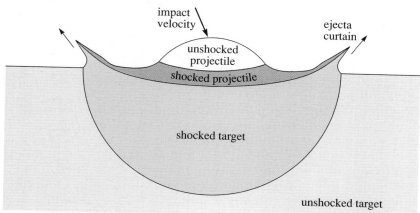

Figure 5.14 An impact seen in cross-section, during the compression stage, a fraction of a second after the arrival of the projectile (which can be anything from about 1 m to 10 km or more in diameter). Melted and fragmented material derived from the shocked regions of the projectile and target (i.e. the surface of the Moon or other body) feeds the beginning of a conical ejecta curtain.

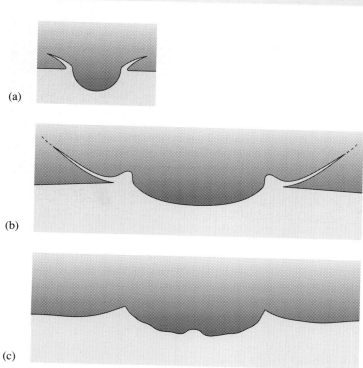

(a)

(b)

(c)

Figure 5.15 Crater formation seen in cross-section (and applicable to all sizes up to many hundreds of km across). (a) Shows an early stage of excavation of the crater, which is widening (but no longer deepening) by the addition of fragmented target material to the expanding ejecta curtain. (b) Drawn at half the scale, this shows the final stage of excavation when the last material joining the ejecta curtain has barely enough energy to flop out over the rim. (c) At the same scale as (b), this shows the eventual form of the crater, after collapse of the walls and growth of a central peak (due to rebound at the site where the projectile struck). The ejecta curtain has by now formed a blanket of ejecta across the surrounding terrain.

The material flung out of a crater while it is being excavated is known as **ejecta**, and on the Moon (and many other airless bodies) this contributes to a blanket of fragmented and powdered debris several metres thick known as the **regolith**. The regolith is the lunar 'soil', in which the *Apollo* astronauts left their footprints (Figure 5.16).

Figure 5.16   An astronaut standing on the lunar regolith, a fragmentary surface composed of ejecta from impacts on a wide variety of scales.

Although radiometric dating of samples that have been returned to Earth now allows us to date many events in lunar history quite precisely, the basic stratigraphy of the near-side of the Moon was worked out on the basis of detailed telescopic observations, in advance of the first direct exploration of the Moon. In 1962, Eugene Shoemaker and Robert Hackmann, of the US Geological Survey, were among the first to demonstrate that the lunar highlands are the lowest stratigraphic unit on the Moon, and that a succession of other rock units overlap onto the highlands, and are superimposed in turn on each other. Clearly, those units that overlie other units must be younger. This progressive superposition of younger units is the basis of the **lunar stratigraphic time-scale**. The younger stratigraphic units on the Moon include vast areas of basaltic lava (the **lunar maria**, from the Latin for 'seas', singular *mare*) and widespread ejecta blankets from major craters. By imagining these younger layers stripped away, it is possible to visualize what the early surface of the Moon was like. This has been done for you in Figure 5.17, which shows the Moon at present, and at two times in the past.

Figure 5.17   The Moon at three times in its history.

(a)   Today, naming several of the maria and showing sites from which samples
have been brought to Earth. *A = Apollo* landing site (US manned exploration); and
L = Luna   landing site (Soviet unmanned exploration).

❑   Compare Figures 5.17b and 5.17c (overleaf). What can you say
about sites where the majority of the maria were emplaced?

■   They appear to have flooded several large circular structures.

This is not mere speculation, because parts of the rims of these large
circular structures remain visible today.

❑   Can you suggest how they were formed?

■   They are generally reckoned to be very large impact craters.

Figure 5.17 (b) The Moon in post-Imbrian times, but before some of the more recent craters were formed. 'Imbrian' is defined in the text.

Craters of this size are usually referred to as **impact basins**. The largest obvious basin lies towards the northwest of Figure 5.17c (it has a prominent slightly younger 200-km-diameter crater within it, to the northwest of its centre), and consists of several concentric rings, the outermost being 1 500 km in diameter. Most of it is now occupied by the lava-filled region that has been known since the 17th century as Mare Imbrium, and so this particular impact basin has been named the **Imbrium basin**. Radiometric dating (of glasses that represent chilled samples of melt produced by each impact) suggests that the major impact basins were excavated during an interval of some 200 million years about 4 000 Ma ago. The Imbrium basin is one of the youngest of these. The lunar stratigraphic time-scale can conveniently be divided into **pre-Imbrian** (all events prior to the formation of the Imbrium basin, >3 900 Ma), **Imbrian** (extending from the formation of the Imbrium basin until the flooding of the basins by mare lavas was complete (3 900–3 200 Ma)) and **post-Imbrian** (<3 200 Ma), the absolute ages being derived radiometrically. Many subdivisions are recognized on the basis of detailed stratigraphy, but they need not concern us in this Course.

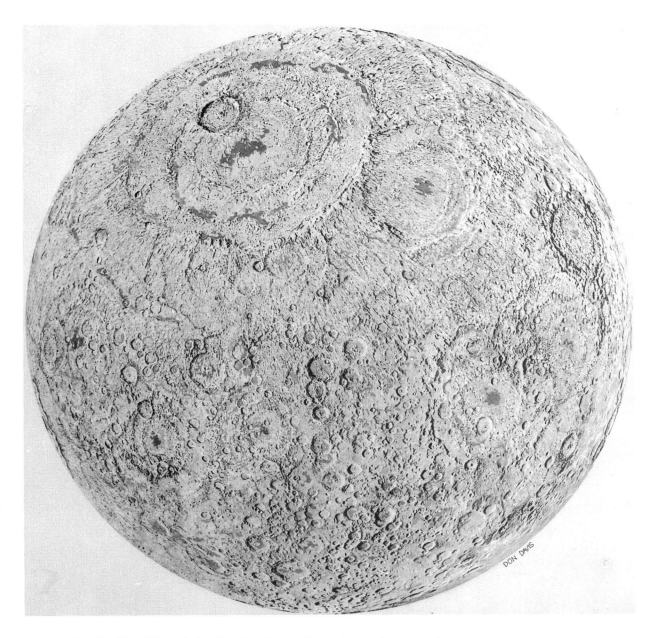

Figure 5.17   (c)   The   Moon during Imbrian times, prior to the emplacement of the maria.

ITQ 9

Figure 5.18 shows a comparison between the lunar highlands (4 420 Ma) and a mare surface that has been radiometrically dated as about 3 500 Ma. Given the respective ages of these two regions, do you think it likely that the rate of impact cratering has remained approximately uniform during the past 4 420 Ma? If not, how would you describe the variation in this rate over time?

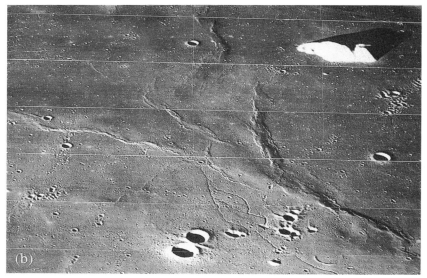

Figure 5.18   (a) A region of the lunar highlands about 300 km across. (b) Part of a lunar mare (oblique view), at about the same scale.

The manner in which the rate of lunar cratering declined from the intense bombardment that it must have experienced in the aftermath of accretion is unclear. Two alternatives are indicated in Figure 5.19, one showing a single steep decline from a sustained heavy bombardment rate experienced by the lunar surface throughout the first 500 million years of its life and the other showing several pulses of cratering prior to the eventual decline. In each case, the mare-filled impact basins and most of the craters visible on the lunar highlands date from immediately before and during the major decline in the rate of cratering at about 3 900 million years, an event that is known, for obvious reasons, as the **late heavy bombardment**. What went on before this is uncertain, and is probably unknowable, because the vast numbers of craters produced during the late heavy bombardment obliterated all the earlier craters. It seems certain, however, that the late heavy bombardment represents the final mopping up of virtually all the debris left within the inner Solar System by the processes of planetesimal accretion and planetary embryo collisions described in Block 1, Section 1.3. Subsequent to the late heavy bombardment, impact cratering has continued (at a much slower rate) as a result of impacts by meteorites, asteroids and the nuclei of comets. Craters produced by these impacts add to the numbers of craters on the highlands, scar the mare surfaces, and puncture the more recent ejecta blankets.

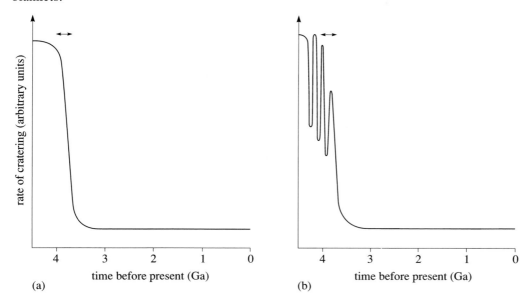

Figure 5.19   Alternative models of the rate of impact cratering on the Moon (and Earth) over time. (a) A simple decline from a peak value. (b) A more cataclysmic view with several pulses of bombardment during the first 500 million years. Arrows indicate duration of the late heavy bombardment. In both examples, the rate of cratering is shown as steady after the end of the late heavy bombardment, although it is not possible to rule out variations, such as brief spasmodic flurries of cratering, or a gradual overall decline.

There is no space to go into crater counting in any detail here, but it is important for you to realize that the more impact craters a surface has, the older it must be. The link between the ages of lunar surfaces and the density of craters superimposed on them is the basis of the **cratering time-scale**, which, having been calibrated radiometrically at several places (the *Apollo* landing sites were chosen to enable each mission to sample units of a variety of ages), can be used with reasonable confidence to infer an absolute age for surfaces that have not been sampled but on which the density of craters can be counted. Furthermore, by counting the abundances of craters with different diameters on younger surfaces (e.g. the maria) and older surfaces (the lunar highlands), one can determine a **size–frequency distribution** of craters of different ages, which shows that the late heavy bombardment produced a greater proportion of large craters relative to small ones than was produced subsequently. This is attributable to there having been a greater proportion of larger bodies in the population of impactors responsible for the late heavy bombardment than in the present suite of impacting bodies. Examples of size–frequency distribution curves are drawn in Figure 5.20. You are not expected to understand how such a plot is constructed, only the significance of what it shows.

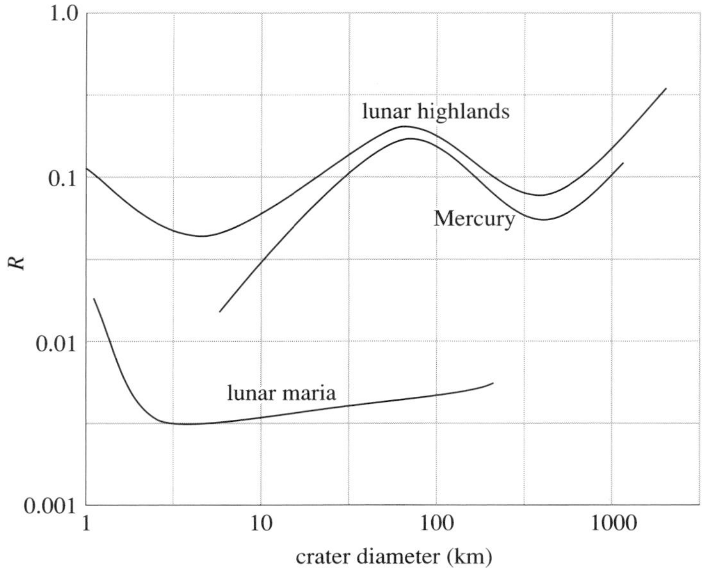

Figure 5.20 A plot comparing the size–frequency distribution of craters on the lunar highlands with that of craters on the lunar maria and on Mercury. Crater diameter is plotted in a logarithmic fashion on the horizontal axis. The vertical axis shows a parameter R (also plotted logarithmically), which is related to the number of craters of each size per km^2; the greater the value of R, the denser the cratering. The shape of each curve is indicative of the size–frequency distribution of the impacting bodies. A comparison between the lunar highland and lunar maria curves shows that the impactors responsible for producing the craters on the lunar maria had a completely different size–frequency distribution to the late heavy bombardment impactors responsible for the craters on the highlands (note that the highlands will include some young craters produced at the same time as those on the maria, but their effect is swamped by the much more abundant older craters). The significance of the curve for Mercury is discussed later.

Now, since the Moon and the Earth orbit the Sun at the same distance, it should be apparent that both must have experienced essentially the same history of bombardment. Thus, the early Earth may once have been as densely cratered as the lunar highlands, depending on the rates at which volcanic, tectonic, erosional and sedimentary activity were able to obliterate the craters. No impact craters dating from the late heavy bombardment can be found on Earth, simply because no undeformed tracts of near-surface crustal rocks of sufficient age have survived. There are impact craters of course, but most are no more than a few hundred million years old (Figure 5.21).

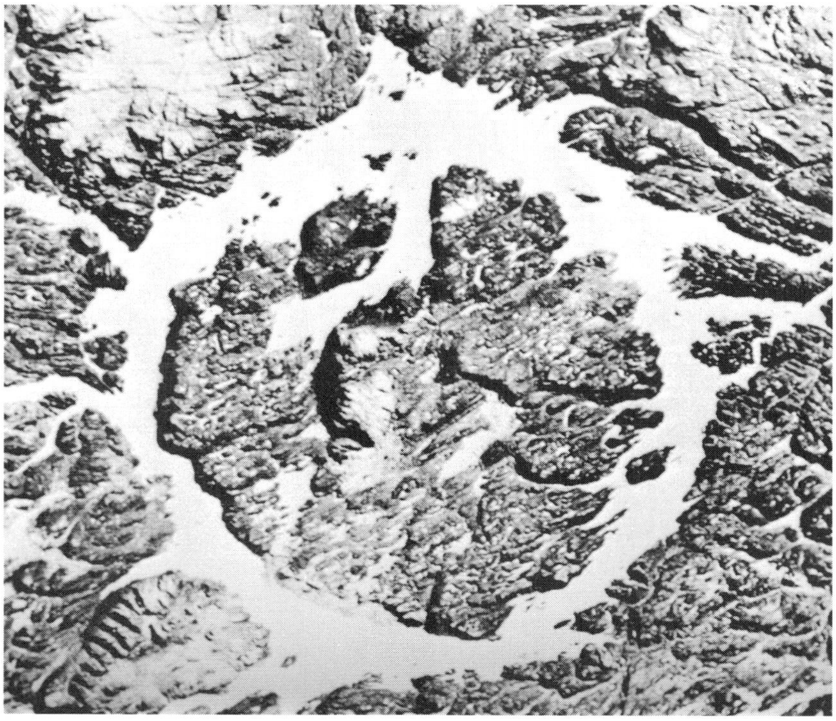

Figure 5.21 A deeply eroded remnant of an impact crater on the Earth, the 214 Ma old *c.* 60 km diameter Manicouagan crater (partly filled by an ice-covered lake) in Quebec, Canada.

ITQ 10

What effect will the late heavy bombardment have had on the geothermal gradient of the primitive Earth?

When we move beyond the Earth–Moon system, we reach bodies whose surfaces have not been dated by independent radiometric means, but on most of which there are large numbers of impact craters. For example, Figure 5.22 shows a view of part of Mercury which is about as densely cratered as the lunar highlands. But does this mean that this surface is the same age as the lunar highlands? Well, until we have samples to date radiometrically, we cannot tell for sure. However, there are reasons for believing that it is. One is that the craters on Mercury have pretty much the same size–frequency distribution as lunar highland craters (Figure 5.20), which shows that both bodies were bombarded by populations of impactors with the same array of sizes. The simplest interpretation is that Mercury was bombarded by the *same* population of impactors as the Moon; in other words, it too experienced the late heavy bombardment. The second reason is that mathematical models suggest that orbital perturbations are likely to have scattered potential impactors fairly evenly throughout the inner Solar System, so the bombardment history of the Moon probably reflects that of Mercury, too. Thus, in default of anything better, the lunar cratering time-scale is applied with reasonable confidence to all the planets from Mars inwards. It is only when we reach the satellites of Jupiter and beyond where we find completely different crater size–frequency distributions (and where theory suggests that different populations of impactors are likely to have been available), that the cratering time-scale breaks down. However, even in these remote reaches of the Solar System, as you will discover in Sections 5.10 and 5.11, the overriding principle that (on any given body) more craters indicates greater age remains by far the best way of deducing relative ages. Before progressing to the next Section, you are advised to listen to AV 10, 'Planetary Matters', which revises (and elaborates on) the fundamental concepts of the cratering time-scale, and introduces you to the nature and significance of several other Solar System bodies.

Figure 5.22 Part of Mercury, as seen by the probe *Mariner 10*. The largest craters in this view are about 70 km in diameter. A scarp running obliquely down this view is one of many on Mercury, which may indicate a period of global contraction (due to cooling) about 3 000 Ma ago.

SUMMARY OF SECTION 5.5

The Earth's surface must once have been as heavily cratered as the lunar highlands, but none of this ancient crust survives in recognizable form. The lunar highlands are anorthositic in composition, and radiometric dating shows that they formed about 4 420 Ma ago. They are interpreted as the Moon's primary crust, anorthite crystals having floated towards the top of the magma ocean.

The rate of impact cratering on the Moon (and by implication all the inner planets) declined to about its present level at about 3 900 Ma ago. During this decline, and immediately prior to it, the remaining debris left over from accretion was mopped up. This episode is known as the late heavy bombardment. Subsequent impact cratering is attributed to meteorites, asteroids and comets, and has continued at a roughly uniform rate. Several large impact basins survive on the Moon, dating from the time of the late heavy bombardment. The largest clear example is the Imbrium basin. Over an interval from about 3 900 to about 3 200 Ma ago, this and most other impact basins were largely flooded by basaltic lavas, forming the lunar mare.

Lunar stratigraphy can, at its most simple, be divided into pre-Imbrian, Imbrian and post-Imbrian. These units were first defined by their superposition relationships and by the densities of superimposed craters, and their ages have since been demonstrated radiometrically (>3 900 Ma, 3 900–3 200 Ma, and <3 200 Ma respectively).

OBJECTIVES FOR SECTION 5.5

When you have completed this Section, you should be able to:

5.1 Recognize and use definitions and applications of the terms printed in the text in bold.

5.8 Understand the principles by which the ages of cratered surfaces on the Moon and elsewhere in the Solar System can be determined, and the limitations of this method.

5.9 Recognize evidence for, and the significance of, the late heavy bombardment.

Apart from Objective 5.1, to which they all relate, the two ITQs in this Section test the Objectives as follows: ITQ 9, Objectives 5.8 and 5.9; ITQ 10, Objective 5.9.

You should now do the following SAQs, which test other aspects of the Objectives.

SAQS FOR SECTION 5.5

SAQ 6 *(Objectives 5.1 and 5.9)*

Explain whether or not the surfaces of the lunar maria record any of the traces of the late heavy bombardment.

SAQ 7 *(Objective 5.8)*

One of the lunar maria has 6 000 craters larger than 1 km diameter per million square kilometres. A region on one of the satellites of Saturn has 3 000 craters larger than 1 km diameter per million square kilometres. Explain whether or not this evidence makes it possible to determine which of the two terrains has the younger surface.

SAQ 8 *(Objectives 5.1 and 5.9)*

During which of the three main lunar stratigraphic time intervals were the lunar highlands formed, and what is the evidence for this?

5.6   HOW OTHER PLANETS WORK

So far in this Course (at least until you reached the previous Section), you have been largely concerned with the Earth itself, to the near exclusion of other planets. This is entirely natural: the Earth, after all, is where we live, and is the planet that we know most about. For geochemical studies, the upper part at least of the Earth's continental crust can be sampled fairly easily (though you should bear in mind that it is impossible to derive a reliable average from such data — Block 1, Section 1.4.2). Samples from the upper mantle and the oceanic crust can be obtained with a little ingenuity. The physical properties of the Earth's interior can be probed by a host of seismic techniques, by gravity surveys, and (although we have not discussed these) by magnetic surveys. Contact with the ground is essential for seismic studies, but large-scale or deep-probing magnetic and gravity work can benefit from the more comprehensive view provided by Earth-orbiting satellites. You should by now have a good grasp of how scientists have sought to determine the internal structure and composition of the Earth, and how these data may be used to assemble a picture of how the Earth works, and how it has evolved from its initial state to what we see (or infer) today. But did the Earth *have* to develop in this way? Are the Earth's structure and present day volcanic and tectonic processes the inevitable results of the evolution of any comparable planetary body? If not, how else might the Earth look today?

To try to answer these questions by mathematical modelling would be horrendously complex, and also untestable. A more promising (and more exciting) approach is to compare the Earth's structure and evolution with that of other bodies in the Solar System. As you saw in AV 10, 'Planetary Matters', the Solar System provides us with a ready-made laboratory, and by looking at how planetary bodies work, in particular at how heat is transferred from their interiors to their surfaces, we can gain an insight into the Earth as well. Note that for the rest of this Block, as in AV 10, we shall be using the term **planetary body** to refer in general to either a planet or a natural satellite of a planet, particularly ones that are comparable in some way with the Earth. The final video for S267, VB 07, 'Other Worlds', is associated with the part of the text you have now reached. We suggest you view this at your convenience, sometime between now and when you complete Section 5.8.

The discussion of heat transfer in the Earth earlier in the Course (particularly in Blocks 1 and 2) can be summarized by saying that most of the transfer of heat from the Earth's mantle to its surface occurs by **plate recycling**, by means of the exposure of hot material at spreading axes and the recycling of cold dense material at subduction zones, and that lesser proportions of heat transfer occur by conduction through the lithosphere and by volcanism at hot spots. The data on other planetary bodies are necessarily much less complete than for the Earth, but this deficiency is partly offset by the large number of bodies we can study. Thus, by way of a finale, we are going to examine some of the other planetary bodies in the Solar System. The study of other worlds is a fascinating discipline in its own right, the capacity of which has increased dramatically over recent years, but we do not have time for more than a general look at some of the more outstanding features. We will narrow the field by concentrating on those bodies that appear to have undergone internal differentiation (into core, mantle and crust) of the kind experienced by the Earth. However, as you saw in Section 5.3, the concept of 'crust' is by no means straightforward. Planetary crusts can be divided into three categories: *primary crust*, representing the earliest stable, solid differentiated outer layer of a planetary body (perhaps formed by fractional crystallization from a magma ocean), of which no remnants survive on Earth; *secondary crust* (the result of partial melting of the mantle), such as the Earth's oceanic crust; and *tertiary crust* (formed by the reprocessing of secondary crustal material), exemplified by the Earth's continental crust. The crust issue aside, we shall be concerned to identify wherever we can an outer lithospheric shell overlying a convecting asthenosphere, because these are likely to be the planetary bodies with tectonic and volcanic processes akin to those occurring on the Earth, which you investigated in Blocks 2, 3 and 4. If you refer back to Block 1, Section 1.7.5, you will find that the definition adopted for the Earth's asthenosphere is that it is coincident with the partially molten low-velocity zone. However, we have little or no seismic data with which to recognize such a layer in other planetary bodies, so in this Block we will relax our definition of asthenosphere to encompass the convecting part of the mantle in general.

ITQ 11

Refer back to Block 1, Figure 1.2 and Table 1.1. Which bodies in this Table are likely to be most Earth-like in their structure?

For the purposes of this Block, where we are concerned with the structure, composition and evolution (but not the orbital characteristics) of planetary bodies, we shall follow the accepted practice among planetary scientists and regard the Moon as one of the **terrestrial planets** (along with Mercury, Venus, Earth and Mars). Having been formed from planetary embryos that accreted in the same region of the solar nebula, their compositions are similar. However, compared with the Sun (Block 1, Sections 1.2 and 1.4), they are lacking in light elements and those which form volatile compounds; notably hydrogen, helium, carbon, nitrogen and neon. Upon differentiation, this recipe is liable to lead to a core composed largely of iron and a little nickel (and possibly a light element), and a mantle and crust dominated by silicates.

❑   Can you recall what the light element(s) in the Earth's core is (are) likely to be?

■   As discussed in Block 1 (Section 1.11.3), the most likely contenders at present are oxygen, or sulphur, or potassium, or carbon, or hydrogen. It is not now thought likely that much silicon is present. You should bear in mind that our evidence for the composition of the Earth's core is circumstantial, and is perforce even more so for the other terrestrial planets!

Depending on their thermal histories, the other terrestrial planets are likely to have now, or to have had when they were younger and hotter, a convecting asthenosphere below their outer, rigid lithosphere, as occurs in the Earth today. We will examine the structures and histories of the Moon, Venus and Mars in Sections 5.7–5.9.

❑   Do you think it would be worth trying to identify lithospheres and asthenospheres in the outer planets, namely Jupiter and the ones beyond it?

■   In general, no.

As you can see in Block 1, Table 1.1, Jupiter, Saturn, Uranus and Neptune are all large bodies of low density. They were formed further from the Sun, from material that condensed at lower temperatures than in the inner Solar System. This gave them a much greater proportion of volatile elements than in the terrestrial planets, which they have been able to retain by virtue of their stronger gravity resulting from their greater masses. The outer layers of these **giant planets** are gaseous (Figure 5.23), and any rocky component they may possess is buried very deeply. If any processes that would be recognizable by a geologist occur within them, they certainly cannot be investigated with our present technology.

Figure 5.23 A view of part of Jupiter seen by the space probe *Voyager 1*, showing violent storms in its hydrogen- and methane-dominated atmosphere.

Possible internal compositions of two of the giant planets are indicated in Figure 5.24. We will not consider these any further. Pluto, the outermost planet, is a different proposition. It is much the least massive of all the planets, despite which it has a higher density than Neptune, the nearest other planet to it. We know little about it at present, but, as you will see shortly, Pluto may be Earth-like in some important attributes. Discounting the planets themselves, there is one more set of bodies in the Solar System where we can study the operation of differentiation, tectonics and volcanism.

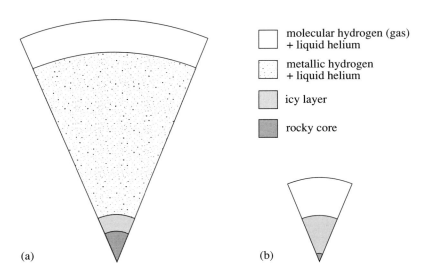

molecular hydrogen (gas)
+ liquid helium

metallic hydrogen
+ liquid helium

icy layer

rocky core

(a)

(b)

Figure 5.24 Possible internal structures of (a) Jupiter and (b) Uranus.

❑   Bearing in mind what you learned in AV 10, 'Planetary Matters', what are these bodies?

■   They are the satellites of the outer planets, many of which have notable similarities to the terrestrial planets in terms of volcanism and inferred internal layering.

Table 5.2 shows some physical data for all the satellites of the outer planets that are big enough to take on a roughly spherical shape, as a result of their own gravity (the limiting size is about 200 km radius). The orbits of most large satellites lie in the plane of their planet's equator and orbital motion is in the same direction as the planet's spin, and so these satellites are thought to have arisen from aggregations of planetesimals within a disc of gas and dust that accompanied each of the giant planets during its formation; a process mimicking, in miniature, the development of the Solar System itself (Block 1, Section 1.3). However, Neptune's large satellite, Triton (Plate 5.1), orbits the planet in the opposite direction to the planetary rotation (in other words, its orbit is retrograde), showing that Triton was almost certainly captured by Neptune some time after its formation. The planet Pluto (Block 1, Table 1.1) seems to be an icy body similar to Triton in mass and composition, and these two bodies may both have formed beyond the orbit of Neptune; however, in terms of their composition and the geological processes that they have experienced, they can both conveniently be grouped with the satellites of the outer planets.

In 1951, G. Kuiper proposed that a zone should exist beyond Neptune's orbit where icy planetesimals that had been unable to assemble into larger planetary embryos should have survived. This Kuiper belt (Block 1, p13) remained hypothetical until its first two members were discovered telescopically in 1992 and 1993, orbiting at about 44 AU and estimated at about 200 km in diameter. These were designated $1992QB_1$ and 1993FW, but their discoverers called them (unofficially) 'Karla' and 'Smiley'. In the following two years a score or so more were discovered, and in 1995 an extrapolation based on a small area of sky surveyed in detail using the Hubble Space Telescope suggested that the total number greater than about 30 km across may be in the region of 10^8. Pluto and Triton may prove in origin to be simply the largest and nearest of the Kuiper belt objects.

Table 5.2 Physical data for planetary satellites. Distances are quoted from the centre of each planet. For satellites from Jupiter to Neptune, the sizes, masses and densities are derived largely from data transmitted to Earth by the space probes *Voyager 1* and *Voyager 2*, that passed through these systems between 1979 and 1989. Satellites smaller than about 200 km in radius are non-spherical, and are either fragments formed by the break-up of larger bodies as a result of collisions or were never part of a larger body in the first place. The two satellites of Mars and some of the smaller satellites of the outer planets are probably captured asteroids.

Planet	Satellite	Mean distance from planet (10^3 km)	Orbital period (Earth days)	Radius (km)	Mass (10^{20} kg)	Mean density (10^3 kg m^{-3})
Earth	Moon	384.4	27.32	1738	734.9	3.34
Mars	Phobos	9.37	0.32	$13.5 \times 10.7 \times 9.6$	1.3×10^{-4}	2.2
	Deimos	23.46	1.26	$7.5 \times 6.0 \times 5.5$	1.8×10^{-4}	1.7
Jupiter	Io	421.6	1.77	1815	894	3.6
	Europa	670.9	3.55	1569	480	3.0
	Ganymede	1070	7.16	2631	1482	1.9
	Callisto	1883	16.69	2400	1077	1.9
	12 others			<135		
Saturn	Mimas	185.5	0.92	197	0.38	1.2
	Enceladus	238.0	1.37	251	0.8	1.2
	Tethys	294.7	1.89	524	7.6	1.3
	Dione	377.4	2.74	559	10.5	1.4
	Rhea	527.0	4.52	764	24.9	1.3
	Titan	1221	15.95	2575	1348	1.9
	Iapetus	3561	79.33	718	18.8	1.2
	10 others			<175		
Uranus	Miranda	129.8	1.41	236	0.8	1.4
	Ariel	191.2	2.52	579	13.5	1.7
	Umbriel	266.0	4.14	586	12.7	1.5
	Titania	435.8	8.71	790	34.8	1.7
	Oberon	582.6	13.46	762	29.2	1.6
	10 others			<85		
Neptune	Triton	354.8	5.88	1350	214	2.1
	7 others			<200		
Pluto	Charon	19.1	6.39	596	11	2.0

ITQ 12

Compare the ranges of size and density of the terrestrial planets (including the Moon) with those of the satellites of the outer planets, using the information in Table 5.2 and Block 1, Table 1.1, and decide which, if any, of these satellites resemble(s) the terrestrial planets in these respects.

You will see in Section 5.10 that Io is a remarkable world. It is essentially composed of silicates, rather like our own Moon, but, for reasons you will discover later, it is the site of a tremendous amount of present-day volcanic activity, whereas the Moon has been essentially inactive for billions of years. Because of its size and composition, it is reasonable to treat Io as another terrestrial planet.

With the debatable exception of Europa, the densities of the other outer planet satellites are too low for them to be silicate bodies, and therefore, from a geological point of view they form a separate (but related) class. Most models of the condensation phase of Solar System formation (Block 1, Section 1.3) predict that beyond a distance of about 5 AU from the Sun the temperature of the solar nebula probably became low enough for ice to condense.

❑   Assuming this is correct, which planets and satellites would have been liable to accrete a significant proportion of water or ice?

■   Block 1, Table 1.1, shows that Jupiter is at 5.2 AU from the Sun whereas Mars is very much closer to the Sun, so it is likely that ice could only become a major component of the Jupiter system and those further from the Sun.

As you know, ice is a lot less dense than rock, and the low densities of most of the satellites of the outer planets can be explained if they are a mixture of these two components. Theory can be backed up by observation in this instance, because there is a clear signal attributable to ice in the spectrum of sunlight reflected from these bodies (Figure 5.25). The lack of ice on Io, which is demonstrable from both its density and its reflectance spectrum, can be explained by inwardly increasing temperatures in the dust and gas cloud around Jupiter at the time when the satellites were being accreted, so that it became too hot for water to condense in the region where Io was forming. Excluding Io, the satellites of the outer planets are sometimes referred to as the **icy satellites**.

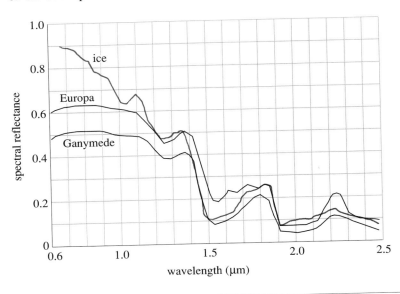

Figure 5.25   The reflectance spectrum of ice compared with spectra of two icy satellites of Jupiter. The vertical axis shows spectral reflectance, which is the proportion of incident sunlight reflected at each wavelength. The satellites are darker than pure ice because of meteoritic dust and other rocky material on the surface, but the characteristic spectral features of ice, notably the absorption (i.e. troughs in the reflectance curves at wavelengths near 1.25, 1.55 and 2. μm, remain clearly visible. Note that visible light has wavelengths between about 0.4 μm (blue) and 0.7 μm (red)

ITQ 13

Given that the density of any silicates within a body such as Enceladus (a satellite of Saturn) is likely to be around 3.0×10^3 kg m^{-3} and that the density of ice is around 0.95×10^3 kg m^{-3}, what percentage (by volume) of rock would account for a body with the same density as Enceladus, the density of which is 1.2×10^3 kg m^{-3} according to Table 5.2, if it were composed solely of rock and ice?

Thus, Enceladus appears to be about seven parts ice and one part rock. The same principle can be used to estimate the rock and ice proportions of any body, though in the case of the larger satellites the values for average density of each component have to be adjusted to allow for increasing density with depth within the body, due to self-compression (Block 1, Section 1.2.2). In reality, things are probably more complicated still, because the rock may be hydrated (which could decrease its density by about 0.5×10^3 kg m^{-3}), and there may also be significant proportions of carbonaceous material instead of rock, or volatiles such as ammonia mixed within the ice. Possible internal structures of icy satellites are shown in Figure 5.26.

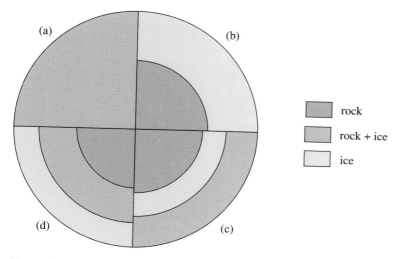

Figure 5.26 Possible internal structures of icy satellites. (a) An undifferentiated ice–rock mixture. (b) Fully differentiated with a rocky core overlain by ice. (c) Partially differentiated with a rocky core and an icy mantle overlain by a homogeneously accreted ice–rock crust that has never melted. (d) Differentiated, with ice–rock overlying a rocky core and with ice forming the outer layer. The degree of differentiation will depend on the extent to which each body accreted homogeneously or heterogeneously, and on its subsequent heating history.

Note that Figure 5.26 shows *compositional* layers only. The surface temperatures on the icy satellites are so low (only 100 K even at Jupiter) that the ice behaves in a rigid fashion. It does not behave at all like glacier ice on Earth, which will flow downhill; instead it is very strong and behaves like the rock forming the Earth's lithosphere (whether defined elastically or thermally; see Block 1, Section 1.13). However, if the temperature rises sufficiently with depth, then the underlying ice becomes weak enough to convect. It is thus decoupled from the overlying lithosphere, and can be regarded as an asthenosphere (irrespective of whether or not it includes a partially molten layer). The small size of the icy satellites (relative to the larger terrestrial planets) facilitates heat loss, but this is compensated by the lower temperatures at which ice is able to convect, with the result that several of them show remarkable signs of geological activity. You will have the opportunity to discover more about some of these bodies in Section 5.11 and in VB 07, 'Other Worlds'.

You may wonder why we are not going to consider the asteroids (Block 1, Section 1.2.1) to help us understand how planetary bodies such as the Earth work. The reason is that they are too small. The largest asteroid has a radius of about only 500 km, and the vast majority are much smaller. If, as seems likely, these have a similar suite of compositions to the meteorites, being essentially stony or nickel iron in composition, then their evolution must have been complete very early in the history of the Solar System. Heat from the decay of ^{26}Al and other short-lived isotopes (Block 1, Section 1.3.3) may have allowed the larger asteroids to differentiate, but as these isotopes are long since vanished there is no likely mechanism of heat generation that could have enabled such comparatively small rocky bodies to generate sufficient internal heat to remain active for more than a few tens of millions of years. In contrast, the icy satellites (although some are no bigger than the largest asteroids) need much less heat to maintain internal convection, and, as you will discover in Sections 5.10 and 5.11 there is an important heat source other than decay of radioactive isotopes that is responsible for long histories of activity on many of them.

SUMMARY OF SECTION 5.6

The major bodies in the Solar System can be divided into the terrestrial planets (Mercury, Venus, Earth, the Moon, and Mars), the giant planets (Jupiter, Saturn, Uranus and Neptune) and the satellites of the outer planets (which Pluto probably resembles). Only the terrestrial planets and some of the planetary satellites are likely to have a recognizable lithosphere and asthenosphere, or to have had such a distinction at some stage in their documentable histories. Most of the largest satellites of the outer planets seem to have formed within a disc of gas and dust that surrounded each planet at the time of its formation.

Differentiation will produce a nickel–iron core and a silicate mantle in the case of the terrestrial planets, and a rocky core with an icy mantle in the case of the icy satellites of the outer planets. Io, the innermost major satellite of Jupiter, resembles a terrestrial planet in terms of its density and mass. The other large satellites of the outer planets have icy surfaces and (except for Europa) significantly lower densities. Icy satellites have an ice lithosphere with a very low surface temperature ($<100\,K$), which will be underlain by an ice asthenosphere, if the internal temperature gradient is sufficient.

OBJECTIVES FOR SECTION 5.6

When you have completed this Section, you should be able to:

5.1   Recognize and use definitions and applications of each of the terms printed in the text in bold.

5.10 Recognize the distinctions between the major bodies of the Solar System, and appreciate the factors influencing their overall composition during condensation from the solar nebula.

5.11 Be able to work out the proportions of rock and ice in a two-component model for the composition of a planetary body, given its density and size, and apply similar logic for any two-component mixture.

Apart from Objective 5.1, to which they all relate, the three ITQs in this Section test the Objectives as follows: ITQs 11 and 12, Objective 5.10; ITQ 13, Objective 5.11.

You should now do the following SAQs, which test other aspects of the Objectives.

SAQS FOR SECTION 5.6

SAQ 9 *(Objectives 5.1 and 5.10)*

Why is it reasonable to classify the Moon as a terrestrial planet, and how does its likely origin differ from that of most of the major satellites of the outer planets?

SAQ 10 *(Objectives 5.10 and 5.11)*

(a)   Look at the physical data for Europa (a satellite of Jupiter) in Table 5.2. Bearing in mind the methodology of ITQ 13, what proportions of ice and rock must be mixed to account for its density?

(b)   Now consider the spectral evidence for the composition of Europa's surface shown Figure 5.25. How can you reconcile Europa's surface composition with its density?

(c)   Assuming the locations of their present orbits are inherited from the time of their formation, can you suggest any genetic link between the compositions and densities of Io, Europa and Ganymede and the regions in which they formed in the gas and dust cloud around the young Jupiter (the proto-jovian nebula)?

5.7 THE MOON

In Block 1, Section 1.3.1, you have already learned something about how the Moon may have formed. You saw in Section 5.5 of this Block that it preserves a record of cratering dating back to the end of the late heavy bombardment, and how this provides the basis of a cratering time-scale that can be applied to other bodies in the Solar System. In this Section, we shall be more concerned with the Moon's interior.

The Moon is the natural starting point for a comparison between the Earth and other planetary bodies, because of its proximity and the variety of ways in which it has been studied. For example, it is the only other planetary body on which humans have landed: on six occasions between 1969 and 1972 as part of the US *Apollo* programme. The Moon is the smallest and least massive of the terrestrial planets (Block 1, Table 1.1), and gives us an idea of the fate to be expected for any Earth-like body within which heat generation becomes insufficient to maintain asthenospheric conditions close enough to the surface to enable volcanic and tectonic processes to be manifested.

Figure 5.17a in Section 5.5 showed the appearance of the Moon, as seen from the Earth (you may like to compare this with what you can see with your own unaided eyes or through binoculars). As you should recall, the Moon's surface has two major divisions; the maria and the highlands. Section 5.5 proposed that the highlands are the primary crust of the Moon (probably produced by fractional crystallization of a global magma ocean) whereas the maria, which have lower crater densities, are younger and occupy regions (including the major ancient impact basins) that were flooded by basaltic lava flows. Geochemical evidence suggests that these **mare basalts** were supplied by partial melting in the mantle.

In considering the evidence we have for the nature of the Moon's interior, it is worth beginning by considering what we can deduce from the relationship of the Moon to the Earth, as determined by simple visual observation of its surface.

ITQ 14

We always see the same pattern of maria when we look at the lunar surface, showing that the Moon always keeps the same face towards the Earth. What does this tell us about the rate at which the Moon rotates about its axis compared to the time it takes to complete an orbit around the Earth?

When a satellite, such as the Moon, rotates once per orbit, it is said to be in **synchronous rotation**. Such a situation is common among planetary satellites in general, and can hardly be due to coincidence. In fact, it is regarded as an inevitable consequence of tidal interactions between a satellite and a considerably larger planet. Just as the Moon raises tides in the oceans but also much smaller ones within the solid Earth, so the Earth raises tidal bulges in the solid Moon (one in the centre of the near-side and, for reasons that we need not go into, a similar one on the opposite hemisphere). The Moon was probably originally rotating much faster than at present. In those days, the Moon's lithosphere would have been continually flexing to allow the tidal bulges raised by the Earth on the Moon to maintain their positions relative to the Earth. The rate of expenditure of energy in this flexing (an effect known as **tidal dissipation**; Block 1, Section 1.3.3), may have contributed significantly to the rate of heat generation within the early Moon, but tidal forces have long since slowed the Moon's rotation to the point where the tidal bulges remain stationary. In a similar manner, dissipation of tidal energy due to the propagation of tidal bulges raised by the Moon on the Earth is gradually slowing the Earth's rotation and contributing very slightly to the Earth's internal heat generation, as noted in Block 1, Section 1.3.3.

5.7.1   HEAT GENERATION IN THE MOON

Because of the Moon's synchronous rotation, the tidal bulges raised by the Earth on the Moon (which distort the Moon less than 1 metre from its equilibrium shape) are effectively fixed in place.

ITQ 15

What is the implication of this for the importance of tidal dissipation as a heat source within the Moon today?

Thus, tidal heating is probably not a major heat source within the Moon, although it may have been important in the distant past, before the Moon's rotation became synchronous.

❑   What then is likely to be the major source of present-day heat production within the Moon?

■   It is reasonable to assume that energy is not being released at a significant rate by processes such as core formation and internal compression. Short-lived radioactive isotopes have long since decayed. This leaves radioactive decay of the long-lived isotopes of U, Th and K.

Our knowledge of the actual rate of lunar heat loss is very limited. However, we have much more experimental data for the Moon than for any planetary body in the Solar System other than the Earth, and so it is worth looking at these data in some detail to assess what they tell us about the Moon's composition and structure. Heat flow was successfully measured at two *Apollo* landing sites, and had a value of $21\,mW\,m^{-2}$ at the *Apollo 15* site and $16\,mW\,m^{-2}$ at the *Apollo 17* site. Because of the non-representative situations of the two landing sites, near the edges of mare basins (see Figure 5.17a), it is far from straightforward to estimate a global average heat flow from these two measurements. Values in the range $11–18\,mW\,m^{-2}$ have been suggested; a value of $15\,mW\,m^{-2}$ would be a reasonable compromise. By contrast, the Earth's present heat flow is about $80\,mW\,m^{-2}$. To keep the following ITQ simple, you are asked to assume that all the measured heat flow (on both bodies) derives from radiogenic sources, and that none of it represents long-term cooling by loss of primordial heat (this is actually a matter of considerable conjecture and debate; Block 1, Section 1.12.1). In order to compare the rates of heat production in the Earth and the Moon, it is necessary to allow for their respective surface areas and masses.

ITQ 16

(a)   The Earth's surface area is approximately $5.1 \times 10^{14}\,m^2$, and that of the Moon is approximately $3.8 \times 10^{13}\,m^2$. Use the values for their average heat flows (per unit area) given above to calculate their total, global, rates of heat production.

(b)   Now divide the Earth's and the Moon's global rates of heat production by their masses (Block 1, Table 1.1) to derive their rates of heat generation per kg.

Your results from these calculations should show you that the rate of heat generation per kg within the Moon is roughly similar to that of the Earth. In fact, if you consider the uncertainty in the derivation of the global average value for lunar heat flow, it should be apparent to you that the lunar heat-flow data provide no grounds for suggesting that the Earth and Moon are significantly different in their rates of heat generation per kg.

❑ If we stay with the assumption that global heat flow is effectively all due to radiogenic sources, does this mean that Moon rock has about the same concentration of radioactive elements as is typical of terrestrial rocks?

■ It is not as simple as that. Remember that U, Th and K are all essentially lithophile elements, so they will occur primarily within the crust and mantle. A substantial fraction of the Earth's mass (about a third) resides in its core, so U, Th and K will tend to be concentrated into the remaining two-thirds (with the possible exception of K, which could be the 'light' element in the Earth's core; see Block 1, Section 1.11.3). We have not yet discussed the Moon's core, but the Moon's low density (see Table 5.2) should make it apparent to you that it is impossible for the Moon to have a substantial nickel–iron core like the Earth (in fact, the Moon's core, if it has one, can represent no more than about 2% of its total mass). Thus, laying aside the possibility of K in the Earth's core, the Earth's radiogenic heat production must be concentrated into the two-thirds of its mass represented by the crust and mantle, whereas the Moon's radiogenic heat production is distributed throughout its whole mass (given that its core is small enough to be ignored). A fairer comparison between heat generation per kg in the Earth and the Moon would be between the Earth's mantle plus crust and the bulk (i.e. whole) Moon. So, the value of the Earth's heat generation per kg calculated in ITQ 16 should be multiplied by 1.5 (i.e. divided by two-thirds) to derive a value of heat generation per kg excluding the core, giving approximately $10.2 \times 10^{-12} \, \text{W kg}^{-1}$. This might suggest that the Earth's mantle plus crust has a greater abundance of heat-producing elements than the Moon, but such a conclusion is largely undermined by the assumptions we have had to make, notably that the value chosen for the lunar global average heat flow is correct, that the heat flow is all attributable to radioactive decay (which is unlikely to be true; Block 1, Section 1.12.1), and that we can ignore K in the Earth's core.

We have now reached an important conclusion. Based on measurements of heat flow and density, all that it is reasonable to say about the Moon's concentration of radioactive elements is that it is unlikely to differ by more than a factor of two from the average for the Earth's crust and mantle. Moreover, this has told us nothing about the *relative* abundances of U, Th and K. You may recall from Block 1, Section 1.12.1 (see also Figure 5.9), that the present-day rate of heat generation by K in chondrites is a little more than that of U and Th combined; however, in a non-chondritic Earth (with K depleted) the contributions of the three elements to heat generation could be about equal.

Fortunately for us, we have more that just heat-flow data available to help us determine the Moon's composition, and we will look at some of the other data next.

5.7.2 THE COMPOSITION OF THE MOON

Lunar samples weighing a total of 382 kg were brought to Earth for analysis from the six sites investigated by the *Apollo* manned lunar landings, and further much smaller samples were obtained from three sites by the Soviet robotic *Luna* probes. A set of samples collected at so few sites is not quite so unrepresentative of the whole lunar surface as you might think, because crater-forming impacts have acted to redistribute material from neighbouring areas in the form of recognizable blocks of ejecta. In addition, the overall chemical composition of large tracts of the lunar surface was mapped by X-ray fluorescence and using gamma-ray spectrometers operating from the *Apollo 15* and *16* Command Modules in orbit, while the more newsworthy Moon-walking activities were going on below. Based on these analyses, it is possible to construct detailed models involving fractional crystallization and partial melting to try to reconstruct the chemistry of the deeper source rocks, and thence of the Moon as a whole. All this forms a considerable body of data and theory, which we do not have time to examine in detail in this Course. The best we can do is to attempt to summarize the main points:

1   The Moon's crust in the highlands has on average 45.0% silica, and is essentially composed of plagioclase (An_{95}–An_{97}) with lesser amounts of low-Ca pyroxene and olivine. This is similar to the uncommon terrestrial rock type called anorthosite, and the lunar highland crust is often referred to as being anorthositic in composition, as you saw in Section 5.5.

2   The maria are composed of silica-poor lavas, generally referred to as mare basalts, although their silica content ranges from 48.8% down to 37.8%, and thus falls partly below the 52–45% range generally adopted for basalts (Block 3, Appendix 2). Moreover, compared to terrestrial basalts they tend to be higher in TiO_2, FeO and MgO, and poorer in Al_2O_3 and alkalis, especially Na_2O.

3   Almost all lunar rocks have no detectable trace of H_2O, and the Moon as a whole appears to be depleted in water, and volatile elements such as K.

4   Conversely, refractory elements, notably U, are enriched in the Moon compared to the Earth. (You should have reached these conclusions about lunar depletion in volatiles and enrichment in refractory elements, relative to the Earth, for yourself in Block 1, ITQ 7.)

ITQ 17

The Moon shows notable characteristics in its siderophile element content as well. Table 5.3 summarizes the elemental abundances in C1 carbonaceous chondrites (Block 1, Figure 1.15) and in recent estimates for the mantle plus crust of the Earth and of the Moon. Examine this Table and decide whether the Moon appears to be enriched or depleted in siderophile elements compared to the Earth.

Thus, the Moon is depleted in volatile elements and siderophile elements, and enriched in refractory elements. Phosphorus, which does not appear in Table 5.3, is both volatile and siderophile, and it should be no surprise to you to learn that its abundance in the Moon is only about one-third of its abundance in the Earth's upper mantle.

These observations are compatible with (though they do not *prove*) the giant impact hypothesis for the Moon's origin that you learned about in Block 1, Section 1.3.1. The depletion of volatile elements in the Moon (relative to the Earth) can be attributed to the high temperatures at which the material that went to form the Moon was vaporized.

Further evidence bearing on the Moon's origin comes from **oxygen isotopes**. Oxygen has three stable isotopes (^{16}O, ^{17}O and ^{18}O), and the ratios between these in lunar materials are indistinguishable from the same ratios in rocks from Earth, whereas chondritic meteorites have a different isotopic signature.

Table 5.3   Oxide and elemental abundances in C1 chondrites, primitive Earth mantle (present mantle plus crust) and the bulk Moon (excluding any core). Some values differ from those appearing in Block 1, either because they are based on averaging different samples, or because of different assumptions in the models.

	C1 chondrite	Earth mantle + crust	Moon mantle + crust
	Major oxides (%)		
SiO_2	34.2	49.9	43.4
TiO_2	0.11	0.16	0.3
Al_2O_3	2.44	3.64	6.0
FeO	35.8	8.0	13.0
MgO	23.7	35.1	32.0
CaO	1.89	2.89	4.5
Na_2O	0.98	0.34	0.09
K_2O	0.10	0.02	0.01
	Volatile elements		
K (ppm)	854	180	83
Rb (ppm)	3.45	0.55	0.28
Cs (ppb)	279	18	12
	Moderately volatile element		
Mn (ppm)	2490	1000	1200
	Moderately refractory element		
Cr (ppm)	3975	3000	4200
	Refractory elements		
Sr (ppm)	11.9	17.8	30
U (ppb)	12.2	18	33
Th (ppb)	43.0	80	112
La (ppb)	367	551	900
Eu (ppb)	87	131	210
V (ppm)	85	128	150
	Siderophile elements		
Ni (ppm)	16 500	2000	400
Ir (ppb)	710	3.2	0.01
Mo (ppb)	1380	59	1.4
Ge (ppm)	48.3	1.2	0.0035

These isotopes do not take part in radioactive decay nor are they products of such a process, so any differences in their relative abundances must be inherited from the solar nebula. The fact that the Earth and the Moon have indistinguishable oxygen isotopic composition is strong evidence that they formed from material that condensed at the same distance from the Sun (another point in favour of the giant impact hypothesis), whereas meteorites reflect a trend in the relative abundances of oxygen isotopes with increasing distance from the Sun.

❑   Can you suggest how the depletion in siderophile elements in the Moon's mantle plus crust (Table 5.3) would be compatible with the giant impact hypothesis if the Moon does in fact have a core (albeit small)?

■   The material making up the Moon would have come from the mantles of the Earth and the impacting body (Block 1, Section

1.3.1), so it would have had low siderophile abundance to begin with. Any episode of core formation in the Moon would have scavenged most of this already-low siderophile content into the Moon's core, leaving the mantle further depleted in these elements.

Earlier (based on the answer to ITQ 16), we showed that heat-flow measurements could be used to suggest that the Moon has a slightly lower abundance of radioactive elements than the average for the Earth's mantle and crust (based on heat generation of 10.2×10^{-12} W kg^{-1} for the Earth and 7.8×10^{-12} W kg^{-1} for the Moon), but we remarked that such a close comparison rests on many untrustworthy assumptions. Subsequent to that, in Table 5.3, we introduced some estimates of the abundance of these elements based on geochemical criteria. In answering the following ITQ, you have an opportunity to see how consistent the two lines of evidence are.

ITQ 18

The rates of radiogenic heat generation from ^{235}U, ^{238}U, ^{232}Th and ^{40}K are quoted in Block 1, Table 1.11 (p.176). Taking this information into account and using the relative abundances of these three elements in the Earth's mantle plus crust and the Moon according to Table 5.3, calculate the resulting rates of radiogenic heat generation per kg in these two planetary bodies. (*Note*: At the present day, uranium consists of 0.7% ^{235}U and 99.3% ^{238}U, thorium is essentially all ^{232}Th , and 0.01% of potassium is ^{40}K.)

Thus, it appears that the rate of radiogenic heat production per kg in the Moon may be about 50% greater than that of the Earth (discounting any K in the Earth's core, which could perhaps restore the balance). This is because the Moon's increased heat production from the refractory elements U and Th (in which it is enriched relative to the Earth) considerably outweighs its reduced heat production from K (in which it is depleted). The value you have derived for the Moon's rate of radiogenic heat production based on the geochemical data in Table 5.3 is satisfyingly close to the 7.8×10^{-9} mW kg^{-1} global heat production extrapolated from heat-flow data that you calculated in ITQ 16. The reason why the Earth's rate of heat loss per kg probably exceeds that of the Moon is that in the Earth radiogenic heat generation is supplemented by heat from other sources, including heat from the Earth's formation (Block 1, Section 1.12.1). Remember though that the elemental abundances in both the Earth and the Moon are uncertain (even for a unit as accessible as the Earth's crust, estimates of the abundance of K differ by a factor of 2 in Block 1, Table 1.3, p. 49!).

The exercise you have just been through is a prime example of the maxim in planetary science that the more data you have, the harder it is to 'prove' that your model is consistent with the 'truth'. Lunar global heat flow extrapolated from a few surface measurements leads to an estimate of the lunar rate of heat generation that is comparable with the rate of heat generation calculated from estimates of the Moon's content of radioactive elements. This means that the interpretations we have drawn from the types of data are mutually consistent, but it does not prove that our compositional model is correct. Even assuming that we have estimated the Moon's global heat flow correctly, other combinations of U, Th and K could be found that would fit the heat flow just as well, and in any case we cannot tell how much, if any, of the heat flow comes from non-radiogenic sources. When we consider other planetary bodies in subsequent Sections, if things seem cut-and-dried to you, try to remember that in these other examples there are fewer data to confuse the issue, so it is easier to set up a model but harder to refute it. However, despite the fact that it is difficult to explain the *details* of the Moon's geochemical structure and evolution, we nonetheless have a vastly better picture of it than we did in the pre-*Apollo* era. To round off our brief investigation of the Moon, we shall now consider the geophysical evidence for the Moon's structure.

5.7.3 THE MOON'S INTERNAL STRUCTURE

Orbital spacecraft have demonstrated that the Moon has an extremely weak magnetic field, with a strength at least 10^7 times less than that of the Earth.

❑ Do you think it likely that the Moon's present magnetic field could be generated by motions within an electrically conducting core, in the manner suggested for the Earth in Block 1, Section 1.11.3?

■ This is very unlikely. The Moon's magnetic field is far too small and, as you have seen, the Moon's density rules out a core of any significant size.

In fact, the Moon's magnetic field is generally attributed to magnetism imposed on tiny metallic iron grains in lunar rocks as they cooled by the presence of a now-vanished magnetic field. Such a field could have been produced by the Moon's small core when it was still hot enough to convect, or by interactions between the Moon and charged particles streaming out from the Sun (the solar wind).

As you should have realized from your study of Blocks 1 and 2, we have probably learned most about the internal structure of the Earth from seismic techniques. We are fortunate that a four-station seismic network was set up on the lunar surface by the *Apollo* programme. This was still functioning when reception of the transmitted data was terminated on 30 September 1977 (for budgetary reasons). The most abundant seismic events recorded were weak moonquakes at depths about half-way between the centre and the surface of the Moon, on the near side, whose occurrence was strongly correlated with tides raised on the Moon by the Earth and the Sun. Shallower moonquakes were much rarer, though stronger, and were independent of tides. The total lunar seismic energy release turned out to be small, less than 10^{11} J per year compared to about 10^{18} J per year for the Earth, and the individual events are small, ranging up to 3.3 on the Richter scale for shallow events and only up to about 1 for deep events.

❑ The seismic stations that were part of the network were set up at the *Apollo 12, 14, 15* and *16* sites. Check where these are on Figure 5.17a. Do you think this is a good distribution for stations intended to study the deep structure of the Moon?

■ It is actually far from ideal. The *Apollo 12* and *14* sites are close together, and with the other two sites they form a roughly equilateral triangle about 1 000 km on a side, near the centre of the Earth-facing hemisphere. We would prefer stations to be more widespread, and to include at least one on the far side (in fact, there have never been any landings, robotic or otherwise, in that hemisphere).

Despite these limitations, the *Apollo* seismic network provided invaluable data. Apart from the internally generated moonquakes, seismic waves were recorded that were clearly caused by meteorite impacts (including some on the far side), and in addition a total of nine expended pieces of *Apollo* spacecraft, in the form of upper stages of the main rocket and lunar module ascent stages, were deliberately crashed onto the Moon, thereby generating seismic shocks at precisely known locations.

The seismic signals of both internal and external origin are of markedly different character to those with which we are most familiar from earthquakes, in that the signal from a single event lasts from about 30 minutes to several hours, compared with typical durations of a few minutes on the Earth. 'Ringing like a bell' is how the Moon was described when this phenomenon was first reported. It can be explained if the Moon has an outer layer some hundreds of metres to a couple of kilometres thick within which seismic waves travel slowly (less than $1 \, km \, s^{-1}$) but bounce back and forth within this layer very efficiently, without losing their strength (i.e. without attenuation). This layer is interpreted as a **megaregolith**, consisting of rock that has been fragmented and redistributed by major meteorite impacts (as opposed to the regolith, that you met in Section 5.5, which is the much more finely broken down material arising from the more common smaller meteorite impacts).

The megaregolith serves to complicate the interpretation of seismic signals on the Moon, but it does not prevent the seismic structure of the deep interior being unravelled. Figure 5.27 shows the velocity structure of the outermost 100 km of the Moon derived from P-wave travel times, for the region between the *Apollo 12* and *14* landing sites. As you can see in Figure 5.17a, this is essentially a mare region, where mare basalts have flooded a basin made in highland crust.

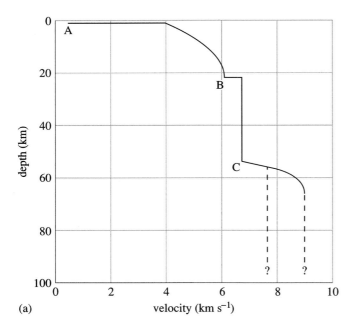

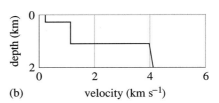

Figure 5.27   (a) P-wave velocity model for the region between the *Apollo 12* and *14* landing sites, derived from arrival-time data. (b) Shows the detail for the first 2 km depth. A–C refer to velocity changes that you are asked to interpret in ITQ 19.

ITQ 19

Examine Figure 5.27 in the light of the previous two paragraphs and attempt to interpret the nature of the material in the following four depth intervals: above A; between A and B; between B and C; below C.

Thus, the lunar crust in this region appears to be about 60 km thick. Near the *Apollo 16* site, it is probably nearer to 75 km thick, and it may be thicker still on the far side. Below the crust is a mantle. Geochemical arguments (similar to those expressed for the Earth in Block 3) suggest that this is peridotitic in composition, and this is compatible with its seismic velocity and the Moon's bulk density.

But what of the deeper interior? One simple but important piece of evidence comes from the focal depths of the moonquakes, which are mostly between 800 and 1 000 km in depth. Shallow moonquakes are much rarer.

❏   How does this compare with the depths of terrestrial earthquakes, and what does it tell us about the thickness of the Moon's lithosphere?

■   Earthquakes are more common at shallower depths (see Block 1, Figure 1.26), and none has been observed deeper than 720 km. Earthquakes originate within the lithosphere. On Earth, the deepest earthquakes occur within descending slabs below subduction zones, though disregarding these the Earth's seismogenic oceanic lithosphere is only about 50 km thick (Block 1, Section 1.13). In contrast, the great depths of most moonquakes shows that the Moon's lithosphere is up to 1 000 km thick (and there is no evidence for subduction zones).

We can get a fuller idea of the structure of the Moon's mantle and its possible division into lithospheric and asthenospheric regions from the travel times of seismic waves originating at events distant from the *Apollo* seismic stations. The inferred velocity structure of the Moon as a whole is presented in Figure 5.28.

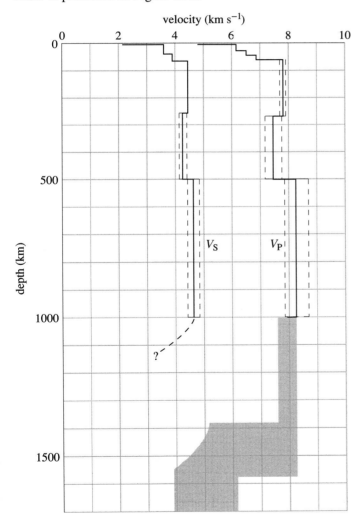

Figure 5.28   P- and S-wave velocity model for the Moon. Velocity changes within the mantle may actually be gradual rather than stepped. At less than 1 000 km depth, the lines shown are the best values with the uncertainties shown either side. Below 1 000 km, P-wave velocities are much more poorly constrained, but probably lie somewhere within the shaded area, and S-waves are strongly attenuated.

❏   Thinking back to the seismic interpretation of the Earth's mantle, can you suggest two possible reasons for the seismic velocity increase at 500 km?

■ This could be attributable to a bulk compositional difference (perhaps the lower mantle is richer in iron) or to a phase change.

These two possibilities are discussed for the Earth in Block 1, Section 1.11.2. Note that taking into account the more gradual rate of increase in pressure with depth within the Moon (a consequence of its smaller mass), the 500 km seismic discontinuity in the Moon cannot represent the same phase change (olivine to spinel, as depth increases) as that proposed for 400 km in the Earth; instead, it may represent a transition from aluminium-rich pyroxene to garnet.

ITQ 20

As indicated in Figure 5.28, below about 1 000 km the P-wave velocity decreases and S-waves are either not transmitted at all or they are weakened so much by attenuation that they cannot be detected. How would you interpret this?

Thus, the propagation of seismic waves indicates that the Moon's asthenosphere probably begins at a depth of about 1 000 km.

❑ How does this tally with the depths of the deep-focus moonquakes?

■ At 800–1 000 km, these appear to be within the base of the lithosphere.

The interpretation below 1 200 km depth in Figure 5.28 is based on a single impact detected from the far side, from which the late arrival of P-waves suggests the possibility of a small, low-velocity core.

❑ Can you recall a non-seismic and non-magnetic geophysical measurement used in Block 1 to help demonstrate the presence of a dense core in the Earth?

■ The technique was based on comparing the Earth's moment of inertia with that of a body of the same mass and size but uniform density (see Block 1, Section 1.10).

The same technique can be applied to the Moon, because its moment of inertia has now been measured quite precisely, by tracking satellites in orbit around the Moon. Three models of the internal density distribution of the Moon, each consistent with its moment of inertia, are shown in Figure 5.29.

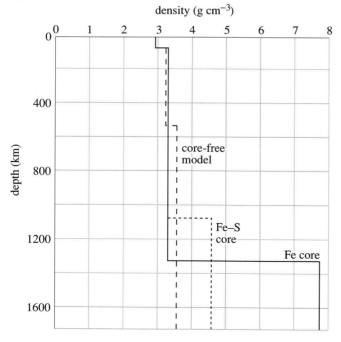

Figure 5.29 Three models for the lunar density distribution based on the Moon's moment of inertia.

❑   Which of these is most compatible with the seismic velocity structure shown in Figure 5.28?

■   The decrease in P-wave velocity at about 1 400 km depth (Figure 5.28) suggests a core with a radius more or less the same as the one in the Fe core model, which is the only one that fits. The Fe–S core is too big (a model with any light element other than sulphur in the core would look much the same), and the core-free model would not result in any P-wave velocity decrease near the centre of the Moon.

Unfortunately, none of this is conclusive. By allowing the density of the middle part of the mantle to be just 100–200 kg m^{-3} denser than shown in the models in Figure 5.29, it is possible to shrink the size of the core down to nothing, or to decrease its density from that of the Fe core model to that of the Fe–S core model and still match both the measured moment of inertia and the seismic data. So, the evidence for a lunar core is inconclusive; the most that can be said is that if one exists then it cannot amount to more than about 2% of the lunar mass and its radius must be less than about 360 km. Our present view of the lunar interior as derived mainly from seismic data and geochemical considerations is summarized in Figure 5.30.

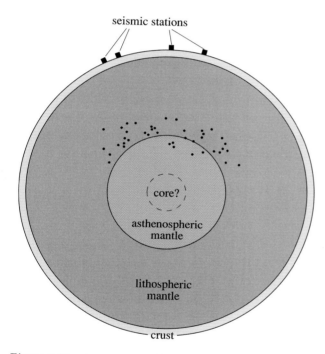

Figure 5.30 A cartoon to illustrate the overall structure of the Moon. The distribution of deep moonquake foci is shown by dots and the locations of the *Apollo* seismic stations are indicated.

There is one further line of evidence about the Moon's interior that we should consider here, and that is the gravity data. Before the *Apollo* landings, several robotic probes were placed in lunar orbit to carry out the essential pre-landing survey work. Apart from recording detailed images of much of the lunar surface, these provided an opportunity to measure variations in the Moon's gravity field by means of precise tracking of their orbital motion. There is no need to go into how this was achieved, beyond remarking that slight mass excesses or deficits below the satellite alter the gravitational force on it, so the satellite's orbit is distorted in response. Some of these early gravity field data are reproduced in Figure 5.31, in the form of a free-air anomaly map (Block 1, Section 1.8.4). This map shows several examples of features that generated much excitement at the time, namely positive gravity anomalies centred over specific sites.

Figure 5.31   A map of variations in the gravity field on the lunar near-side, derived from *Lunar Orbiter V* tracking data. The contour interval is 0.1 Gal. Closed gravity lows are distinguished by shading. Other closed contours indicate positive gravity anomalies.

❑   Does a positive free-air gravity anomaly imply a mass deficit or a mass excess?

■   The fact that the gravitational field at a positive anomaly is stronger than that over the surrounding region shows that there must be a local mass excess.

These positive anomalies came to be known as **mascons**, which is short for mass concentrations.

ITQ 21

Study the distribution of mascons in Figure 5.31 and compare this map with the views of the Moon in Figure 5.17. With which major surface features do the mascons coincide?

In the early days, it was fashionable to attribute mascons to the remains of the large high-density meteorites whose impact formed the respective mare basins. However, more detailed analysis shows that the gravity anomaly representing a mascon is too broad to be modelled by the effective point source that would be formed by such an impactor. Nowadays, the mascons are interpreted as the signature of crustal thinning and associated rapid rebound of the mantle at the times of the basin-forming impacts, exaggerated by the presence of the high-density basalts that now fill the mare basins.

ITQ 22

Whatever their origin, the presence of mascons has important geological implications. What does their existence tell you about whether or not the Moon is in isostatic equilibrium, and how does this relate to the thickness of the lithosphere as shown in Figure 5.30?

An important outcome of lunar gravity studies is that the younger features of the Moon's crust, such as the mascons and the few large post-Imbrian structures, are out of isostatic equilibrium. Some but not all features of Imbrian age are similarly out of isostatic equilibrium, but the older, pre-Imbrian highland topography leaves little or no trace in the free-air gravity field, and has therefore been more or less isostatically compensated.

❏   How would you explain this in terms of cooling and thickening of the Moon's lithosphere over time?

■   The lithosphere must originally have been thin enough to allow isostatic compensation, but by around about Imbrian times it had thickened, as a result of global cooling, to such an extent that it became too thick and strong for any new crustal anomalies to be compensated by differential subsidence.

SUMMARY OF SECTION 5.7

The Moon can be regarded as the smallest and least dense of the terrestrial planets. Its surface is dominated by impact craters, with the youngest widespread volcanism being the infilling of the mare basins that ended about 3.2 billion years ago (defined as the end of the Imbrian period). The highlands represent the early lunar crust, and probably crystallized out of a magma ocean. The Moon is enriched in refractory elements and depleted in both volatile and siderophile elements, compared to the Earth's mantle plus crust.

Heat-flow data suggest that the Moon has a similar rate of heat generation per kg to the Earth, despite the Moon's apparent enrichment in the heat-producing elements U and Th. It is possible that (assuming the heat-flow data are trustworthy) more of the measured heat flow in the Earth comes from non-radiogenic sources than in the Moon.

The Moon's magnetic field is at least seven orders of magnitude weaker than the Earth's. The annual rate of energy release by seismic activity in the Moon is also some seven orders of magnitude less than that of the Earth. Seismic data show that the Moon's lithosphere is about 1 000 km thick, overlying a possibly partially molten asthenosphere.

The younger crustal features of the Moon are associated with free-air gravity anomalies, whereas the oldest are isostatically compensated, showing that the lithosphere has been too thick to allow isostatic compensation since about Imbrian times (c. 3.5 billion years ago). The Moon's lithosphere has probably thickened with time as a result of cooling. The Moon may have a core of either iron or iron and sulphur, but if so it is certainly less than about 400 km in radius.

OBJECTIVES FOR SECTION 5.7

When you have completed this Section, you should be able to:

5.1 Recognize and use definitions and applications of each of the terms printed in the text in bold.

5.12 Account for the Moon's heat flow and composition, and show how the latter is consistent with the giant impact hypothesis of lunar origin.

5.13 Describe the likely internal structure of the Moon, and cite some of the relevant evidence.

5.14 Describe *very simply* the Moon's geological history, in particular the development of its crust and lithosphere.

Apart from Objective 5.1, to which they all relate, the nine ITQs in this Section test the Objectives as follows: ITQs 14–18, Objective 5.12; ITQs 19–21, Objective 5.13; ITQ 22, Objective 5.14.

You should now do the following SAQs, which test other aspects of the Objectives.

SAQS FOR SECTION 5.7

SAQ 11 (*Objective 5.14*)

In the case of the Moon, what is (a) its primary crust, and (b) its secondary crust?

SAQ 12 (*Objective 5.13*)

What is the evidence that the Moon has a partially molten asthenosphere, and how deep is it?

SAQ 13 (*Objective 5.13*)

As shown in Figure 5.30, the *Apollo* seismic network did not detect any deep moonquakes originating on the far side. Bearing in mind the previous discussion about the Moon's internal structure, can you suggest three reasons that could explain this?

SAQ 14 (*Objective 5.13*)

(a) If total lunar seismic energy release is 10^{11} J per year compared to 10^{18} J per year for the Earth, show that the rate of seismic energy release in the Moon is seven orders of magnitude less than that in the Earth.

(b) To what do you attribute the comparative lack of seismic activity in the Moon? (There are several factors to consider.)

SAQ 15 (*Objective 5.14*)

(a) Figure 5.32 is a plot showing the abundances (relative to chondrites) of members of a generally incompatible group of trace elements known as the rare earth elements, arranged in order of increasing atomic number from left to right, in the lunar highland crust and in the inferred source region for the mare basalts. Rare earth elements are extremely useful geochemical tracers, though we have not introduced them until now in S267*. Among these elements, europium (Eu) behaves anomalously in that it can take a divalent form (Eu^{2+}), which is the right size to substitute for Ca in the crystalline lattice of feldspar (unlike the others which are trivalent and are therefore highly incompatible). Can you explain why the otherwise smooth trend in abundances of the rare earth elements is interrupted at europium by a peak (a positive europium anomaly) in the lunar highland crust and a trough (a negative europium anomaly) in the mare basalt source region, in a way that is consistent with the composition and presumed origin of the lunar highlands?

(b) Thinking back to Block 3, Section 3.7.1, would you expect the Rb/Sr ratio to be higher in the lunar highlands or the mare basalt source region?

---

* Exceptionally, in Block 1, Figure 1.5.

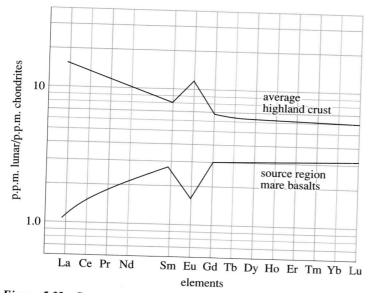

Figure 5.32   Rare earth element abundance pattern for the average lunar highland crust compared with that of the source region for the mare basalts. Note the logarithmic concentration scale. For use with SAQ 15.

SAQ 16 (*Objective 5.12*)

Compared to the Earth's mantle plus crust, the Moon is enriched in refractory elements and depleted in both volatile and siderophile elements (Table 5.3). How may this be explained by the giant impact hypothesis of lunar origin, and subsequent events? (Assume that both the target body (Earth) and the incoming planetary embryo were already differentiated.)

5.8 VENUS

If you consult Block 1, Table 1.1, you will find that of all the planets, only Venus comes within about 5% of the Earth's density and radius. If we want to know whether all 'Earth-like' planets inevitably look like the Earth, then Venus, seemingly being a twin of the Earth, is the obvious example to study. Until recently, we could discover little about the surface of Venus because it is permanently veiled in a thick cloudy atmosphere that prevents us seeing the surface, except with the aid of radar to which the atmosphere is transparent.

Venus's atmosphere is one of many remarkable things about the planet, having a surface pressure about 90 times that at sea-level on Earth. It is about 96% carbon dioxide and 3.5% nitrogen, with traces at around a hundred ppm of water vapour, sulphur dioxide, argon and a few other gases. Through a combination of trapping incoming solar radiation and inhibiting the heat flow from the planet's surface from being radiated to space, this atmosphere makes the surface of Venus about 500 K warmer than it would otherwise be, giving it an average surface temperature of about 730 K (457 °C). This so-called **greenhouse effect** is far greater at Venus than is the case for the Earth, where the atmosphere raises the surface temperature by only about 35 K.

5.8.1 THE COMPOSITION OF VENUS'S SURFACE ROCKS

Our direct knowledge of the surface conditions on Venus comes from soft landings made by Soviet robotic spacecraft (*Venera 8–10, 13* and *14* and *Vega 1* and *2*), between 1972 and 1985. Some of these sent back pictures of the surface around the landing sites, such as those shown in Figure 5.33, and all performed simple chemical analyses of various kinds. The abundances of K, U and Th (determined by *Venera 8–10* and *Vega 1* by means of natural gamma radiation) appear to be more similar to terrestrial rocks than to those found in lunar rocks and meteorites. *Venera 13* and *14* and *Vega 2* carried X-ray fluorescence instruments that were able to make somewhat more complete analyses of the surface chemistry. These results are summarized in Table 5.4.

Table 5.4 The composition of Venus surface rocks at *Venera 13* and *14* and *Vega 2* landing sites, compared with the Earth's crust. ± values show the degree of uncertainty in the measurements.

Constituent	Venera 13	Venera 14	Vega 2	Oceanic basalt (Earth)	Average continental crust (Earth)
SiO_2	45 ±3	49 ±4	46 ±3	51.5	63.3
TiO_2	1.6 ±0.5	1.2 ±0.4	0.2 ±0.1	1.5	0.6
Al_2O_3	16 ±3	18 ±3	16 ±2	16.5	16.0
FeO	9 ± 3	9 ± 2	8 ± 1	12.2	3.5
MnO	0.2 ±0.1	0.16 ± 0.08	0.14 ± 0.12	0.26	0.08
MgO	11 ±6	8 ± 3	11.5 ± 4	7.7	2.2
CaO	7 ± 1	10 ± 1	7.5 ± 1	9.4	4.1
K_2O	4 ± 0.6	0.2 ± 0.1	0.1 ± 0.08	1.0	2.9
SO_3	1.6 ±1	0.9 ±0.8	4.7 ± 1.5		
Total (%)	95.4	96.5	93.5	99.9	92.7

ITQ 23

Examine the data presented in Table 5.4 for the composition of surface rocks at the *Venera 13* and *14* & *Vega 2* landing sites. Are these rocks more similar to oceanic basalt or continental crust on Earth?

This conclusion is compatible with the nature of the surface as seen at the lander sites (Figure 5.33). The only additional direct information we have

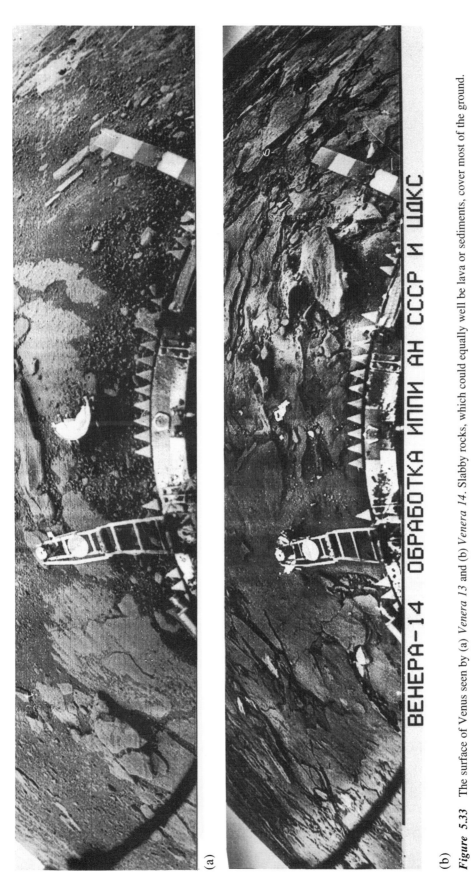

ВЕНЕРА-14 ОБРАБОТКА ИППИ АН СССР И ЦДКС

Figure 5.33   The surface of Venus seen by (a) *Venera 13* and (b) *Venera 14*. Slabby rocks, which could equally well be lava or sediments, cover most of the ground.

(a)

(b)

comes from a single density measurement made by the *Venera 10* lander of $2.8 \pm 0.1 \times 10^3$ kg m^{-3}; this is typical of silicate rocks in general, but is not precise enough to discriminate between rock types, bearing in mind that the surface density may be affected by weathering or fracturing, and the presence of gas bubbles (vesicles) in volcanic rocks.

5.8.2 THE GLOBAL TOPOGRAPHY OF VENUS

The first information about the surface of Venus on a global scale came from an American probe, *Pioneer–Venus*, that went into orbit around the planet and collected data about the surface heights on Venus between December 1978 and July 1980. The way this was done was to measure the distance between the spacecraft and the surface by timing how long it took for radar signals to be reflected back by the surface, in a technique known as **radar altimetry**. The American *Magellan* spacecraft that went into orbit around Venus in 1990 yielded more detailed topographic data, as shown in Plate 5.2

The Venus topography shown in Plate 5.2 makes Venus look very like the Earth. It would be tempting to regard the blue-coded low-lying areas as waterless ocean basins and the green- and yellow-coded areas as continents. As you know, on the Earth the oceans are underlain by a very different type of crust to the continents; they are low-lying due to isostatic forces because the oceanic crust is denser and thinner than that of the continents. This results in a strongly bimodal distribution of topographic heights. You can see this in Figure 5.34, in which the percentages of the Earth's surface area, in 1 km height intervals, are plotted against height relative to the median height for the whole Earth. This is known as a **hypsographic plot**.

Figure 5.34 A hypsographic plot showing the percentage area of the solid surface of the Earth at 1 km height intervals relative to the median height for the entire globe. (Note that the *median* height of the surface of a planet is defined as the height at which 50% of the surface is at a greater elevation and 50% of the surface is at a lower elevation. Also, by convention, high altitudes are shown to the left and low altitudes to the right.) In ITQ 24, you are asked to plot the equivalent data for Venus on this diagram.

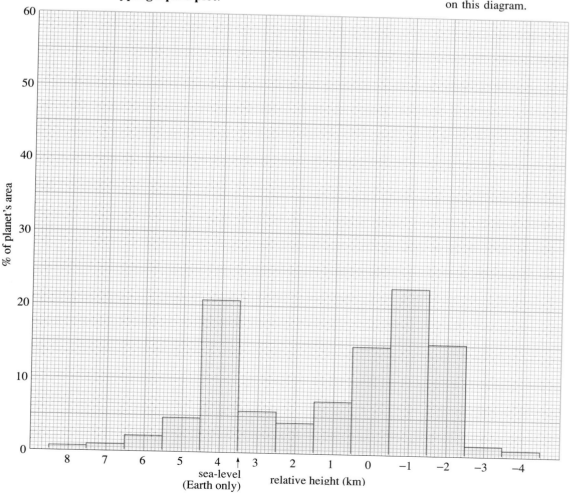

❑   What do you think each of the peaks in Figure 5.34 represents?

■   The peak at a low relative height (i.e. the one to the *right*) is the floor of the oceans (which are typically at 3–6 km below sea-level), and the peak at higher relative height represents the surface of the continental crust (which is mostly less than 1 km above sea-level).

ITQ 24

(a)   Height data for Venus are presented in Table 5.5. Use this information to draw a hypsographic plot for Venus on Figure 5.34, using the same axes as for the Earth.

(b)   What does your hypsographic plot for Venus tell you about the nature of Venus's crust?

Table 5.5   Height data for Venus (obtained by radar altimetry from orbiting space probes) expressed relative to the median height of the surface (which corresponds to a radius of 6 051 km).

Height interval (km) relative to planetary median	% Area in height interval
>5.5	0.04
+5.5 to +4.5	0.1
+4.5 to +3.5	0.4
+3.5 to +2.5	2.7
+2.5 to +1.5	7.8
+1.5 to +0.5	26.4
+0.5 to −0.5	54.2
−0.5 to −1.5	6.6
−1.5 to −2.5	0.01

Thus, there is fairly good evidence by way of the unimodal hypsographic plot that the crustal geophysics of Venus is quite *unlike* that of the Earth. If you add to this the information that the *Venera 14* landing site was about 1 km above the median altitude and that the *Venera 13* and *Vega 2* sites were 1 km higher still, then you can see that there are indications that basaltic crust occurs even at relatively high altitudes. Does this then mean that the crust of Venus has been produced in the same way as the Earth's oceanic crust?

A great deal of caution needs to be exercised in answering this question, not least because the spatial **resolution** of radar altimeter maps (corresponding to the size of smallest features that can be properly seen) is quite coarse. The example in Plate 5.2 is constructed from altimetric measurements with 'footprints' varying in size from about 10–30 km across. This resolution is too coarse to show features such as oceanic spreading ridges in any detail (though the data do appear to rule out major globe-encircling ridge systems comparable to the Earth's Mid-Atlantic Ridge), and would generally not reveal the topographic expression of narrow features such as oceanic fracture zones. The curved troughs visible in the eastern part of Aphrodite Terra are similar in form and scale to trenches associated with island arcs on Earth (Block 2, Section 2.3.3), but this would be slender evidence upon which to interpret them as such.

5.8.3   IMAGES OF THE SURFACE OF VENUS

Another way of using radar to study Venus produces results that are much more spectacular to look at than the altimeter data. This is **imaging radar**, which, as its name implies, produces actual images of the surface. The practicalities of how this is done are beyond the scope of this Course;

essentially, it involves directing a radar signal obliquely onto the surface, and requires complex processing of the returned echoes to construct an image. On the resulting image, light and dark do not correspond to the tones that would appear on an ordinary photograph; instead, rough surfaces appear brighter than smooth surfaces, and slopes facing towards the radar beam appear brighter than slopes facing away. For the purposes of this Course, none of this matters, and you should regard radar images just as you would any other picture of a planetary surface.

Radar images can be made of restricted parts of the surface of Venus using large radio telescopes on Earth, in which case the best spatial resolution achievable is about 1.5 km (Figure 5.35), and the Soviet orbiters *Venera 15* and *16* that arrived at Venus in 1983 produced radar images of much of the northern hemisphere with a resolution of 1–2 km. *Magellan* began orbital mapping in 1990, using an imaging radar the best images from which have a resolution as small as 120 m. Over 99% of the surface was imaged, and examples are reproduced in Figures 5.36–5.40 and Plates 5.3–5.5. You should examine these before continuing with the text.

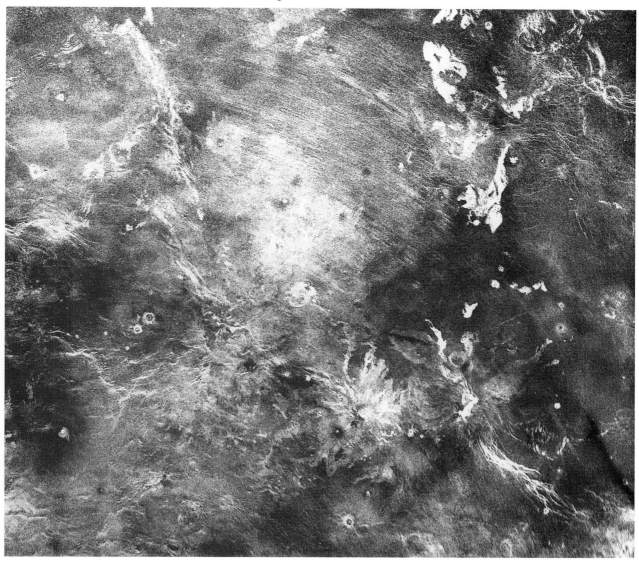

Figure 5.35 An Earth-based radar image of a lowland part of Venus lying between the highlands of Ishtar Terra and Aphrodite Terra, covering 12°–44° N, and 310°–10° longitude (an area greater than China), obtained by ground-based radar. Both impact craters and lava flows can be seen.

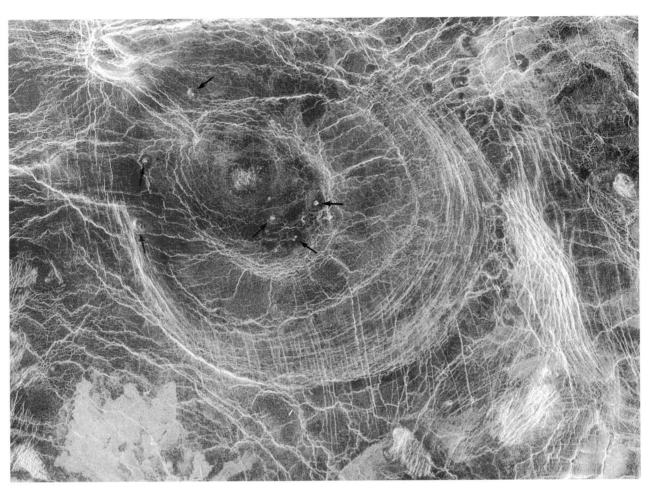

Figure 5.36   A *Magellan* radar image covering one of the many coronae or ring structures on Venus. This one is a gently updomed region of concentric fractures about 230 km in diameter, lying at about 40° N, 20° longitude. The arrows indicate some of the several individual volcanoes superimposed on it.

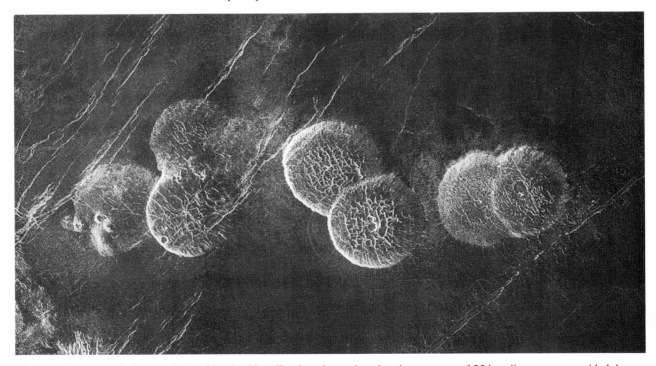

Figure 5.37   A detailed view obtained by the *Magellan* imaging radar, showing a group of 25 km diameter steep-sided domes, each of which probably formed when viscous lavas were extruded. These provide indirect evidence that not everything on Venus is made of basalt, because a basalt would not be viscous enough to produce this landform, which is reminiscent of (generally smaller) terrestrial features formed by the extrusion of rocks containing 60–70% SiO_2.

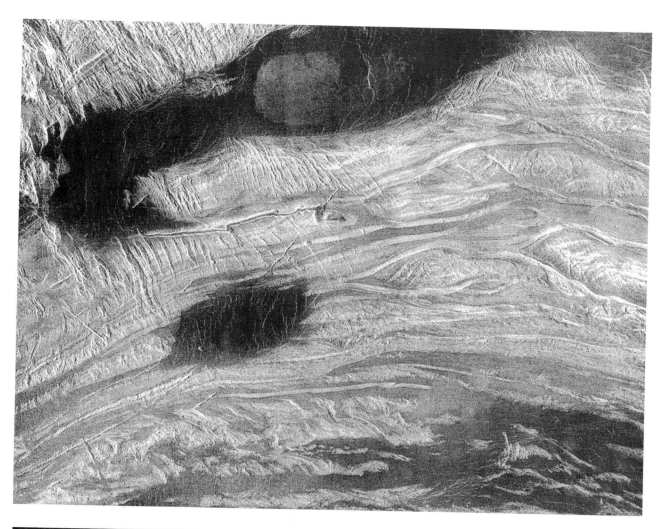

Figure 5.38 A *Magellan* radar image of a 300-km-wide area on the equator near 80° longitude, lying within the highlands of Aphrodite Terra. Radar-bright features pick out a pattern of east–west-trending ridges and valleys reminiscent of fold mountains on Earth, and therefore indicative of north–south compression. Subsequently, a series of apparently tectonic troughs has developed, cutting across the earlier fabric of the terrain, suggesting a later episode of east–west extension. The dark patches are probably areas recently flooded by lavas, giving them a level surface.

Figure 5.39 A 400-km-wide region in the lowlands of Venus near 27° S, 339° longitude bearing three large impact craters. The plains between them are cut by fractures and (barely visible at this scale) volcanic features.

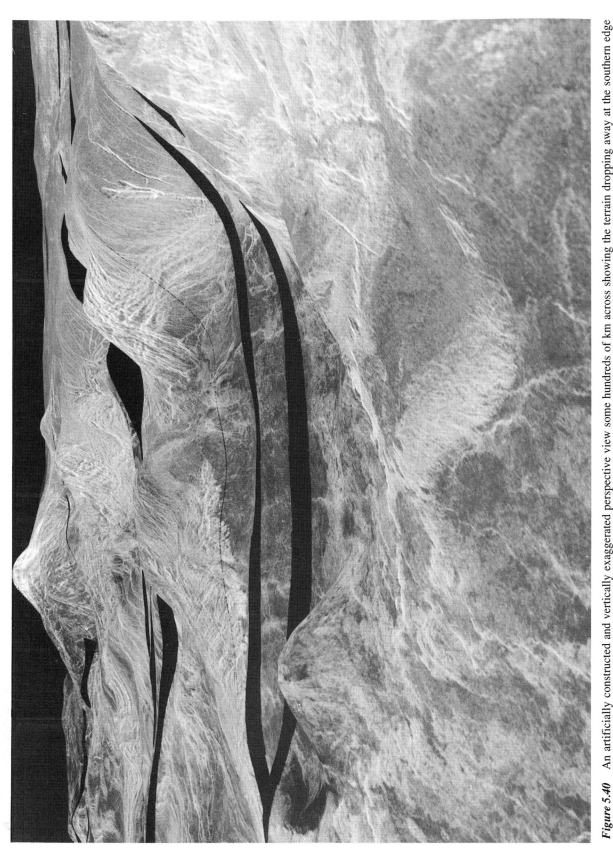

Figure 5.40 An artificially constructed and vertically exaggerated perspective view some hundreds of km across showing the terrain dropping away at the southern edge of Ishtar Terra. The surface details were mapped by the *Magellan* imaging radar, and they have been superimposed on exaggerated topographic data from the *Pioneer–Venus* altimeter. The black strips are gaps in the coverage. The peak in the centre left is probably a basaltic shield volcano similar to Sif Mons (Plate 5.4).

Do not imagine that in these images you have seen examples of every type of important feature on Venus. You may like to pause to consider how representative an overview of the Earth's volcanic and tectonic features you could compile with less than ten pictures! There are further examples of *Magellan* images in VB 07, 'Other Worlds', which includes a visit to one of the laboratories working on the data during the first year of the mission. If you have not yet viewed this video, now would be a good opportunity to do so. Our concept of Venus will undoubtedly evolve as the radar images and the altimetric and gravity-field data from *Magellan* are analysed over the coming years. In the meantime, the nature of Venus's principal surface features can be summarized as follows:

1   Venus has a mixture of highland and lowland terrains, with a similar overall range in height to that shown by the solid surface of the Earth; however, Venus has no clear distinction into oceanic and continental crust.

2   There are abundant volcanic features, including plains apparently covered in basaltic lava. Most volcanoes (ranging from several hundred kilometres down to about 1 km in diameter) have gentle slopes and are consistent in form with basalt volcanoes on Earth. Some steep-sided domes a few kilometres across appear to be constructed of more viscous, presumably more silica-rich, lava (e.g. Figure 5.37).

3   Many volcanic features are concentrated on roughly circular rises about 1 km in height, and surrounded by concentric ridges and fractures (e.g. Figure 5.36). These are known as **coronae** (singular: *corona*, which is Latin for 'crown' or 'halo'), and probably represent sites of persistent mantle plumes. They vary from 75 km to over 2 000 km across.

4   Tectonic features consist of apparently extensional zones and apparently compressional zones (e.g. Figure 5.38). However, there is as yet no convincing proof that there are spreading axes or that 'fold' belts such as in parts of Aphrodite Terra (Figure 5.38) or the edges of Ishtar Terra (Figure 5.40) are associated with subduction zones, and there are few, if any, long strike–slip features with major offsets like the Clipperton Fault (Block 2, Section 2.3.2) or the San Andreas Fault (Block 2, Section 2.5.5) on Earth. Apart from the coronae, deformation is distributed across broad zones one to a few hundred kilometres wide, separating less deformed blocks of hundreds of kilometres width.

ITQ 25

From the description of tectonic features on Venus given above and the images in Figures 5.35–5.40 and Plates 5.3–5.5, how closely would you say that the style of deformation of Venus resembles that produced by plate tectonics in (i) oceanic and (ii) continental regions on Earth?

One factor that could explain the lack of obvious plate-tectonic features on Venus is that the high surface temperature allows the lithosphere to remain warm and buoyant, unlike on Earth where it cools and sinks leading to the slab-pull force that is probably the major plate-driving mechanism (Block 2, Sections 2.6.2 and 2.6.3). Another is that Venus appears to have only one type of crust, whereas subduction on Earth is encouraged by denser oceanic plates descending below less-dense continental plates. An important implication of the apparent lack of plate tectonics on Venus, is that (in contrast to the Earth) plate recycling cannot be the dominant mechanism of heat transfer from the mantle to the surface, and this must be accomplished by a combination of conduction through the lithosphere and volcanism.

Apart from volcanic and tectonic features, there are several major impact craters visible on the *Magellan* images (e.g. Figure 5.39), but these are far rarer than on the lunar maria.

ITQ 26

What does this tell you about the age of Venus's surface?

An element of caution should be injected here, because wind-blown sand may be capable, over long time periods, of eroding or burying craters to the point of invisibility, although this is very unlikely to be capable of destroying craters more than a few km across. Thus, along with the abundance of well-preserved volcanic features and the apparent fold belts and tectonic fractures, the low number of impact craters on Venus is another piece of evidence that (compared to the Moon) Venus has a generally young and geologically active surface. However, unless the age of the surface has been greatly overestimated, the rate of volcanic activity is almost certainly too low to account for the majority of the planetary heat transfer to the surface, and is probably responsible for no more than a few percent of the total. The present evidence does not demonstrate conclusively that volcanic and tectonic activity continues today, but there is no reason to suspect otherwise.

5.8.4   THE INTERIOR OF VENUS

Venus is clearly a differentiated planet. Its density and the few surface analyses that we have (Table 5.4) tell us that. In this Section, we shall consider how much we can deduce about the interior of Venus. We will begin by looking at some of the corollaries of Venus's high surface temperature. Some of the high-resolution *Magellan* radar images show narrow sinuous channels (mostly tens to hundreds of kilometres long, though the longest extends for 6 800 km, making it longer than the Nile), which clearly cannot have been cut by flowing water at the temperatures prevailing at Venus's surface today. The only explanation which seems reasonable is that these channels were cut by flowing lava, presumably one with a low viscosity and a high liquidus temperature, such as an alkali-rich basic melt or an ultrabasic lava like a komatiite (Section 5.4.2). You may have noted in answering ITQ 23 that surface analyses on Venus do appear to reveal komatiitic affinities.

Extensive erosion by lava is rare on the Earth, but it is reasonable to suppose that it could happen more easily on Venus. The high temperature prevailing at Venus's surface suggests two possible reasons for this. One is that because the surface rocks on Venus are already very warm before the arrival of any lava flow on top of them, they are readily melted and incorporated into the overlying flow. The other is that a lava flow might be expected to cool more slowly on Venus, and thus be active for a longer period.

Thus, Venus's high surface temperature goes at least some way to explaining the presence of lava-cut channels. A more fundamental consequence of Venus's high surface temperature is assessed in ITQ 27.

ITQ 27

Let's make the not-unreasonable assumption that Venus's lower crust is anhydrous and similar to granulite in composition, like that of the Earth (Block 1, Section 1.11.1). We can now determine the maximum possible thickness of such a crust, using the fact that this is limited by the depth where anhydrous granulite would begin to melt. This is at a temperature of about 1 100 °C (there is a small influence on this temperature due to pressure, but in a crude model like this we can ignore it). Calculate the depth to this temperature (hence the maximum possible depth of the base of an anhydrous granulitic crust) on (a) Earth, and (b) Venus, taking values for the surface temperatures of the Earth and Venus to be 20 °C and 457 °C respectively. For the purposes of this exercise, you should assume the same average geothermal gradient of 25 °C km^{-1} on both planets.

Thus, any regions of granulitic crust on Venus are unlikely to be much more than half the thickness of the crust on the Earth, if the assumption about comparable geothermal gradients is valid. This assumption is certainly open to dispute, but the important point is that Venus's high surface temperature is likely to have a major influence on the chemical differentiation of its lithosphere, and that this should apply whether the crust is granitic, basaltic or something in between.

❑   What effect do you think Venus's high surface temperature will have on the thickness of the *lithosphere*?

■   Whether we define the lithosphere elastically, seismically, seismogenically, or thermally, or in any other way, it seems likely that the higher surface temperature on Venus would tend to bring the lithosphere–asthenosphere boundary closer to the surface, giving Venus a thinner lithosphere than the Earth. How valid this simple guess is will become apparent in the rest of this Section.

We will now examine some of the evidence and inference relating to Venus's lithosphere, to see how it compares with that of the Earth. Alternative definitions of the term lithosphere were discussed in Block 1, Section 1.13, and you would be well advised to revise that brief Section now, before proceeding.

The transition from lithospheric to asthenospheric conditions within the Earth is influenced by the presence of partial melts, contributing to the low velocity zone or LVZ (Block 1, Sections 1.7.5 and 1.13), and depends also on the properties of the minerals that are stable under the local conditions. Both these factors depend critically on the amount of water present, and water is of crucial importance in determining how the crust of a terrestrial planet evolves. This should be apparent to you (in the case of the Earth) from much of the discussion in Block 4. Unfortunately, there are few direct data bearing on the abundance of water in Venus (e.g. there are no water analyses in the lander data in Table 5.4). In consequence of the lack of these and other data, the nature and thickness of both Venus's crust and its lithosphere have been the subject of much poorly constrained speculation. All we can hope to achieve within this Course is to give you a feel for some of the arguments.

Certainly the surface of Venus is dry, yet if Venus was formed from planetary embryos that grew in the same region of the solar nebula as those that formed the Earth then we might expect the two planets to have similar abundances of water. So, where is the water on Venus? On Earth, most of the water that resides in the oceans, and the bulk of the present atmosphere, is thought to have been released by degassing, in the form of volcanic activity after the last planetary embryo collision. The surface of Venus is too hot for liquid water, so any oceans that Venus ever had would have evaporated. However, there is far too little water vapour in the present atmosphere to account for the disappearance of oceans in this way, so where has the water gone?

There are two totally different ways to explain this. One is to accept that Venus did once have water at the surface in quantities comparable to the Earth, but that it escaped through Venus's atmosphere. This could have taken place as a consequence of intense ultraviolet sunlight in the upper atmosphere breaking water molecules down into hydrogen and oxygen. Hydrogen molecules, being of very low mass, then escaped into space, whereas the oxygen took part in weathering reactions to oxidize the surface rocks. The alternative way to explain the lack of water in Venus's atmosphere is that water was never degassed in major quantities in the first place.

Evidence in favour of the latter hypothesis is provided by the ratios between the isotopes ^{36}Ar, ^{38}Ar and ^{40}Ar in Venus's atmosphere. Compared to the other isotopes of argon, Venus has only one-quarter the proportion of ^{40}Ar prevailing in the Earth's atmosphere. Now, ^{40}Ar (but *not* the other isotopes) is a product of radioactive decay within potassium-bearing minerals, and it is believed that Venus has a potassium content not greatly different from that of the Earth.

❏   What does the relative lack of ^{40}Ar in Venus's atmosphere imply about degassing on Venus compared with the Earth?

■   Being a product of radioactive decay within rocks, the ^{40}Ar in Venus's atmosphere must have been released from rocks by degassing. The low proportion of ^{40}Ar compared to non-radiogenic isotopes of argon in the atmosphere indicates that Venus has degassed less than the Earth.

The conclusion we are forced to is that the present lack of water in Venus's surface environment certainly does not mean that Venus as a whole must have less water than the Earth. In fact, if Venus has degassed less (as indicated by the ^{40}Ar data) but was formed from planetary embryos with the same abundance of water as those that formed the Earth, then it ought to have retained *more* water in its interior than the Earth.

There is some circumstantial evidence that Venus does contain more water than the Earth, which is that volatile inert gases are more abundant in Venus's atmosphere than in the Earth's. This implies that at least the later planetary embryo collisions onto Venus were smaller and less energetic than the Moon-forming impact on the Earth, so that Venus lost less of its volatiles (notably water and these inert gases) than did the Earth.

Therefore, it looks as if Venus's mantle probably contains more water than the Earth's. On Earth, the presence of water is a stimulus to volcanic activity, in settings such as destructive plate margins (Block 4, Section 4.2.1). However, this water is carried down into the upper mantle by descending slabs, in a manner that causes it to be recycled. On Venus, where plate tectonics appears not to occur, the comparatively large amount of water in the mantle seems to lead to a generally *reduced* level of volcanic activity compared to the Earth.

A possible resolution to this apparent paradox is illustrated in Figure 5.41. You have seen this sort of diagram plotting a 'geotherm' and a peridotite solidus before (e.g. Block 3, Figure 3.13). In this example, a solidus has been added that is appropriate to the wet conditions proposed for Venus's mantle. We know from xenoliths that the Earth's mantle is characterized by the presence of anhydrous minerals (Block 3, Section 3.2.1), and so the minerals that crystallize from basaltic partial melts derived from the mantle are also anhydrous. However, a peridotite mantle containing water (such as we might now envisage for Venus) would produce a water-bearing partial melt, and hence hydrous minerals would occur in rocks crystallizing from this melt. Two hydrous minerals are important in this example: phlogopite (a variety of mica) which is stable at pressure above about 3×10^9 N m^{-2}, and hornblende which is stable at lower pressures. Hornblende melts at a higher temperature than phlogopite, and this results in the kink in the wet solidus plotted in the Figure. It is this kink that may be the key to understanding Venus's upper mantle.

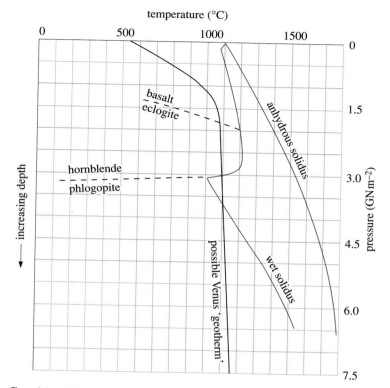

Figure 5.41 Solidus for peridotite under dry conditions (anhydrous solidus) and in the presence of water (wet solidus), with a possible Venus 'geotherm' added. The basalt–eclogite transition is shown, and also the stability fields of hornblende and phlogopite, two water-bearing minerals that occur in eclogite at lower and higher pressure respectively. Note that a pressure of $3.0\,GN\,m^{-2}$ corresponds to a depth of about 100 km within the Earth and 105 km within Venus.

Consider wet mantle upwelling by solid-state convection along the 'geotherm' shown in Figure 5.41. When this rises to a depth equivalent to a pressure of about $3.7\,GN\,m^{-2}$, it crosses the wet solidus and so a basaltic partial melt will form, which will incorporate virtually all the water present. However, at a depth equivalent to about $3.0\,GN\,m^{-2}$, the solidus is recrossed, so no melt is able to reach the surface.

❑　As what sort of rock would this melt crystallize?

■　It would be a hornblende-bearing eclogite, because the point at which the solidus is recrossed lies within the stability field of hornblende and below the basalt–eclogite transition.

It so happens that this eclogite would probably be slightly denser than the surrounding mantle, and would tend to sink and eventually be remelted. Thus, under the conditions reflected by Figure 5.41, melts would be inhibited from approaching the surface, crustal growth would be inhibited and volcanism would be uncommon. The volcanic provinces on Venus such as the coronae are likely to represent areas where the geothermal gradient is different, presumably lying some way to the right of the 'geotherm' plotted in this Figure. Such a situation might be brought about by a mantle plume, or hot spot, comparable to those on Earth discussed in Block 3, Section 3.5.3.

We do not have space here to pursue the complexities of mantle melting in Venus. Instead, we shall confine ourselves to a limited discussion of the global significance of Figure 5.41.

ITQ 28

By analogy with Block 3, Figure 3.31, what is the implied thickness of the lithosphere in Figure 5.41?

The evidence you used in the previous ITQ suggests that the conductively defined (thermal) lithosphere on Venus is substantially thinner than in most places on the Earth, but you should bear in mind that the 'geotherm' in Figure 5.41 is nothing more than a reasonable model; it is not backed up by direct measurements to substantiate it.

❑   On Earth, you know that the weakest part of the asthenosphere is the low-velocity zone (LVZ) that corresponds approximately with the top of the asthenosphere (Block 1, Section 1.7.5). On the basis of Figure 5.41, can you identify the location of a possible LVZ on Venus, and does it directly underlie the lithosphere, as on Earth?

■   Any LVZ would have to be in the region of possible partial melting, where the 'geotherm' falls to the right of the wet solidus, in this case between 3.0×10^9 N m^{-2} and 3.7×10^9 N m^{-2}, which means a depth range of 105–130 km. This is well below the base of the thermal lithosphere as identified in ITQ 28. However, in the absence of seismic data from Venus, the thickness of its seismic lithosphere has not been determined, and it is strictly the base of the seismic lithosphere that is marked by the LVZ on Earth.

Therefore Venus probably does not have a LVZ comparable to the Earth's, because the mantle directly below the thermal lithosphere is not partially molten, and so the upper part of Venus's asthenosphere must be stronger than the Earth's. In fact, it should be stronger than the rest of the Earth's asthenosphere, too. This is because the effect of trapping water in the hornblende-bearing eclogite as implied by Figure 5.41 would be to make Venus's uppermost mantle very dry, in fact considerably drier than the Earth's upper mantle into which water is continuously recycled at subduction zones (Block 4, Figure 4.2), even though below about 105 km Venus's mantle is probably wetter than the Earth's. The extreme dryness of Venus's uppermost mantle will make it more rigid than that of the Earth, even within the convecting sub-lithospheric region.

The essentially geochemical arguments that Venus has a stronger asthenosphere than the Earth are supported by the geophysical data. In particular, there is a strong correlation between Venus's gravity field and its topography, with large-wavelength positive free-air gravity anomalies coinciding with regional topographic highs. This is in marked contrast to the Earth, where gravity is correlated only poorly with regional scale topography (Block 1, Section 1.8.3). The scale of the anomalies on Venus implies that the average depth at which isostatic compensation occurs (which probably gives a reasonable estimate of the thickness of the elastic lithosphere — Block 1, Sections 1.8.5 and 1.13) is generally around 100–150 km, well below the thermal lithosphere, and at the approximate depth of the weakest part of the asthenosphere as suggested by the modelled depth to partial melting. Note that on Earth the depth to compensation is generally less than this, at least in the oceans, and seems to lie at or above the LVZ.

Just to recap on the rather difficult text since ITQ 27: if we interpret the absence of water from Venus's surface and atmosphere to signify that Venus has not degassed very much and that therefore its deep mantle is wetter than the Earth's, *and* if we accept the validity of the phase diagram in Figure 5.41 and the admittedly speculative Venus 'geotherm' shown on it, then the thermal lithosphere of Venus ought to be *thinner* than the Earth's even under the oceans. However, the same data suggest that if Venus has a LVZ then it is deeper than in the Earth, implying that Venus's seismic lithosphere is *thicker* than the Earth's. Independent gravity data suggest that Venus's elastic lithosphere is also thicker than the Earth's oceanic equivalent, and is comparable in thickness with the Earth's continental lithosphere.

In view of these and other complexities and uncertainties, it should come as no surprise that there is considerable debate as to which, if any, of the highland regions on Venus are supported over mantle plumes (hot spots) and which, if any, are thickened regions of crust and lithosphere such as might overlie convective downwellings. Evidence for the former interpretation comes from the abundance of lava flows and other volcanic features on highlands, whereas the latter interpretation is favoured by the apparently compressional tectonics around the margins of some of the highlands. Two competing models are illustrated in Figure 5.42.

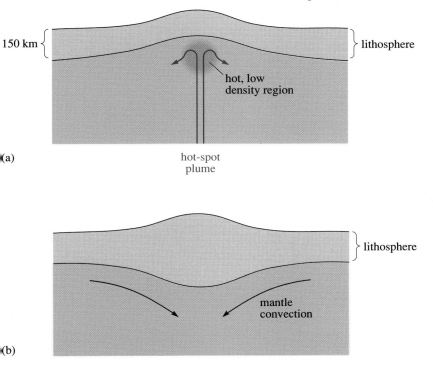

Figure 5.42 Two contrasting models for the isostatic support of highland regions on Venus. (a) Support over a region of hot mantle above a hot-spot plume, leading to thinning of the lithosphere and volcanism. (b) Support by lithospheric thickening over converging convection currents in the underlying mantle.

5.8.5 VENUS'S CORE

Present models for the internal structure of Venus are limited by the total lack of seismic control, and have to be based on consideration of Venus's density and moment of inertia. If you refer back to Block 1, Table 1.1, you will find that Venus is slightly less dense than the Earth. Allowing for the fact that Venus is less massive and that therefore self-compression (Block 1, Section 1.2.2) will be less, the corrected density of Venus works out at about 2% less than the Earth's. There are several factors that could account for this slight difference. One is that the different internal temperature structures of the two planets will cause mantle phase transitions at different depths (e.g. higher temperatures would drive the basalt–eclogite transition to a greater depth in Venus than in the Earth, therefore lowering Venus's density). Others are that Venus could have less core-forming iron and nickel than the Earth, or a greater proportion of a low-density ingredient in its core than the Earth (Block 1, Section 1.11.3). Whatever the subtleties of its exact composition, it seems impossible to envisage Venus without a dense core, dominantly of iron and nickel.

❏ Apart from seismic data and considerations of density and moment of inertia, what geophysical property of the Earth was used to deduce the nature of its core in Block 1, Section 1.11.3?

A fluid, electrically conducting core is indicated by the Earth's magnetic field.

In fact, Venus's magnetic field is at least 25 000 times weaker than the Earth's, but this cannot be used as evidence against the existence of a core. The reason for this is that Venus rotates very slowly, once every 243 days (longer than one Venus year of 225 Earth days; see Block 1, Table 1.1), so any system of rotationally driven currents in its core would be far weaker than in the Earth.

Taking all the evidence into consideration, a feasible core size for Venus in comparison with the other terrestrial planets is given in Figure 5.43. It shows a core consisting of iron, nickel and (possibly) a light element such as sulphur or oxygen (Block 1, Section 1.11.3), the radius of which is a little less than that of the Earth's outer core. This is presumably fluid, but we have no direct evidence, and likewise the presence of a solid inner core can only be guessed at. As you saw in Section 5.8.4, although Venus's crust is probably thinner than the Earth's continental crust, due to the difficulty in wet melts segregating towards the surface, the lithosphere as a whole may be thicker (except on a strictly thermal definition) and more rigid due to the dryness of the uppermost mantle.

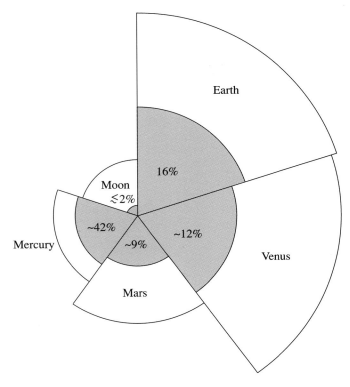

Figure 5.43 Comparison of terrestrial planet size and core radii, showing also the percentage of the total planetary volume occupied by the core.

SUMMARY OF SECTION 5.8

Venus is the most similar planet to the Earth, in terms of size, mass and density, but it has a very high surface temperature (about 457 °C). This probably results in a thinner crust and a thinner thermal lithosphere than in the Earth. Venus's hypsographic plot indicates that the planet has no Earth-like distinction between continental and oceanic crust.

In the few locations where surface rocks on Venus have been analysed, they appear to be essentially basaltic in composition. Radar images show a variety of volcanic features on Venus, most of which are strongly suggestive of the low viscosity typical of basalts, but there are a few examples of more viscous and more explosive features implying that more silica-rich magmas have been erupted in certain places. Many volcanoes are concentrated on coronae, which probably represent sites of persistent mantle plumes.

Apart from the volcanic features, the radar images show impact craters and terrains attributable to both extensional and compressional tectonics. However, there are no unambiguous indications of Earth-style plate tectonics. Based on crater counting, the average age of Venus's surface is around 500 Ma. Not having suffered a Moon-forming giant impact, Venus may have retained more water than the Earth in its interior. It can be argued that the hydrous nature of Venus's deep mantle inhibits the rise of melts and that its uppermost mantle is drier than the Earth's. Possible consequences of this are: (i) if Venus has a LVZ (corresponding with the base of the seismic lithosphere), then it is likely to begin at a depth of around 110 km; (ii) volcanism on Venus is likely to be largely confined to hot spots above mantle plumes; and (iii) its sub-lithospheric mantle (asthenosphere) is probably more rigid than the Earth's.

There is a strong positive correlation between free-air gravity and topography on Venus, suggesting isostatic compensation at a depth of about 100–150 km (which may indicate the base of the elastic lithosphere). However, it is debatable whether the highlands are supported by warm, less dense mantle above rising plumes, or by thickened keels due to compressional tectonics associated with converging convection cells in the mantle.

Factors that may contribute to the lack of plate tectonics on Venus include: the high surface temperature making the lithosphere warm and buoyant, so there are no cold, dense, descending slabs, and hence a much-weakened slab-pull force; the lack of contrast between denser oceanic plates and less-dense continental plates leading to a similar lack of slab-pull force; and the greater strength of Venus's asthenosphere compared to that of the Earth. In the apparent absence of plate tectonics, and with a probably fairly low degree of volcanic activity, outward heat transfer from Venus's mantle to its surface must be accomplished mostly by conduction through the lithosphere.

Venus probably has an Earth-like core, but as a consequence of Venus's slow rotation it does not produce a strong magnetic field.

OBJECTIVES FOR SECTION 5.8

When you have completed this Section, you should be able to:

5.1 Recognize and use definitions and applications of each of the terms printed in the text in bold.

5.15 Describe the *evidence* bearing on the nature of Venus's crust, mantle and lithosphere and how they compare with the Earth's.

5.16 Discuss the possible internal structure of Venus, and cite some of the relevant evidence.

5.17 Make a qualitative comparison of the relative importances of conduction, plate recycling, and hot-spot volcanism in transferring heat from Venus's interior to the surface.

Apart from Objective 5.1, to which they all relate, the six ITQs in this Section test the Objectives as follows: ITQs 23–26, Objective 5.15; ITQs 27 and 28, Objective 5.16; ITQ 25, Objective 5.17 (in addition to Objective 5.15).

You should now do the following SAQs, which test other aspects of the Objectives.

SAQs for Section 5.8

SAQ 17 *(Objective 5.15)*

What are the implications of the following for the relative ease by which subduction processes could occur on Venus, compared with the Earth?

(a)   Venus's hypsographic plot.

(b)   Venus's high surface temperature.

SAQ 18 *(Objective 5.15)*

In what way is the available evidence for Venus's internal structure inferior to that for the Moon?

SAQ 19 *(Objective 5.17)*

Attempt to rank the following three processes in order of increasing importance as mechanisms of outward heat transfer from the mantle to the surface of Venus, justifying your choice: plate recycling; conduction through the lithosphere; hot-spot volcanism.

SAQ 20 *(Objectives 5.15 and 5.16)*

If Venus has more water in its interior than the Earth, why does Venus have so little water in: (a) its atmosphere; (b) its uppermost mantle?

5.9 MARS

Mars is the next planet out from the Sun beyond the Earth. As Block 1, Table 1.1, shows, its physical parameters lie between those of the Earth and Venus, on the one hand, and the Moon on the other. Mars's lower density than the Earth and Venus is undoubtedly due at least in part to a lesser degree of self-compression because of its lower mass (Block 1, ITQ 4). However, it has been suggested that the low density is exaggerated because Mars accreted from material that had condensed sufficiently far from the Sun that the iron in that part of the solar nebula was oxidized to FeS and FeO, so that its core is likely to consist mainly of these lower density compounds rather than metallic iron. Unfortunately, the available moment of inertia data constrain the radius of the core rather poorly.

❏ If Mars's core is largely FeS, would you expect Mars to have a strong magnetic field, similar to the Earth's?

■ An Earth-like magnetic field requires a fluid, electrically conducting core in a rotating body. Although Mars's rotates about as rapidly as the Earth (Block 1, Table 1.1), an FeS core would be non-conducting even if it were fluid (which is unlikely at its pressure and probable temperature). Mars's magnetic field is, in fact, at least 5 000 times weaker than the Earth's.

The surface of Mars is easier to study than that of Venus, in that Mars's atmosphere is tenuous (with a surface pressure less than a hundredth of the Earth's) and has no perennial cloud cover. Several space probes have visited Mars, notably *Viking 1* and *2*, that included automated landers. As well as sending back pictures from the surface (Figure 5.44), the landers analysed the martian dust, which appears to be dominated by oxidized and hydrated clay minerals, and made an inconclusive search for signs of life by checking for signs of metabolic activity within the dust.

Figure 5.44 The surface of Mars seen by the *Viking 1* lander, showing rocks and wind-blown sand. The largest boulder is about 2 m across.

Because the martian soil is chemically weathered, it is an unreliable guide to crustal composition, still less as a means of assessing Mars's bulk composition. Fortunately, we have a few much fresher rock samples from

Mars available for study in the laboratory in the form of **S N C meteorites**, a group recognized during the 1980s (on a variety of grounds, including their much younger ages than other meteorites and distinctive nitrogen isotope ratios) as being samples of Mars rock that must have been thrown into space as ejecta from impacts. The SNC meteorites are medium- to coarse-grained basic and ultrabasic rocks, and geochemical modelling based on these suggests a bulk composition for Mars given in Table 5.6.

Table 5.6   Modelled composition of Mars based on SNC meteorites.

Mantle + crust (%)		Core (%)	
SiO_2	44.4	Fe	77.8
TiO_2	0.14	Ni	7.6
Al_2O_3	3.02	Co	0.4
FeO	17.9	S	14.2
MgO	30.2		
CaO	2.45		
Na_2O	0.50		
P_2O_5	0.16		
K (ppm)	305		
Cr (ppm)	5400		
Ni (ppm)	400		

❑   What would you say is the biggest difference between the mantle plus crust of Mars (according to the model in Table 5.6) and that of the Earth (Table 5.3, p. 49)?

■   Mars appears to have a substantially greater proportion of FeO than the Earth.

This is compatible with the suggestion earlier that Mars formed from more highly oxidized material than did the Earth. Oxidized iron is enriched in the mantle plus crust (a possible reason why Mars's core is proportionally smaller than the Earth's), whereas iron oxidized by sulphur appears to be concentrated in the core.

5.9.1   THE TOPOGRAPHY AND TERRAIN UNITS OF MARS

Many books have been written about the geology of Mars, but all we have space for here is to summarize the main evidence relating to the gross structure of Mars as a planet. This means we will have to pass over such fascinating topics as the canyons and valley systems on Mars (Figure 5.45) that provide convincing evidence of past climates when (unlike today) water evidently existed as a fluid at the surface. We will begin instead with a brief look at Mars's hypsographic curve, which is drawn in Figure 5.46.

ITQ 29

Examine the hypsographic plot for Mars in Figure 5.46 and compare it with the curves for the Earth and Venus (cf. ITQ 24). Which, if any, of these does the curve for Mars resemble? What are the differences?

Hypsographic plots such as these do not give the complete story. For one thing, they will not show the complete range of heights if the tops of mountains and the floors of trenches occupy too small a fraction of the total area to register. In fact, Mars has a total topographic range of some 31 km, in contrast to a maximum relief of 20 km on the Earth and about 14 km on Venus.

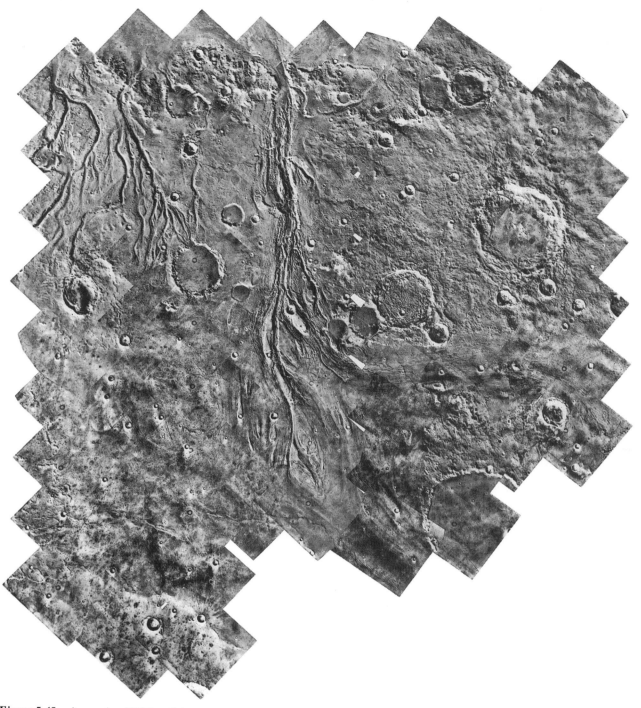

Figure 5.45 A mosaic of *Viking Orbiter* images about 300 km across, showing valley systems that demonstrate the activity of water as an erosive agent in the distant past.

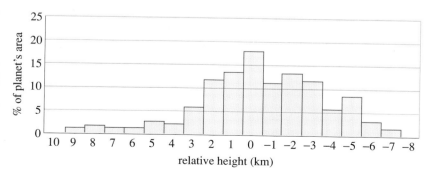

Figure 5.46 The hypsographic curve for Mars. *Note*: elsewhere you may find the 'zero' altitude datum of Mars defined as the altitude where the atmospheric pressure is 611 N m^{-2} (the pressure at which the triple point of water occurs; Block 3, Section 3.3.1), which is *c.* 2.5 km below the median height.

❑   Can you think of four different ways in which areas above the planetary median height on Mars (or any other planet) could be supported?

■   High areas could be supported dynamically by rising mantle plumes, or be thickened by compression over converging convection currents in the mantle (both are suggested for Venus in Figure 5.42), or be supported mechanically by a strong, thick, lithosphere, or be composed of a less-dense type of crust than the lower areas (analogous to continental crust on Earth).

In view of Mars's small size and presumably lower global heat flow than the Earth and Venus, it is to be expected that Mars has a comparatively thick lithosphere, and the third of these explanations is probably the closest to the truth.

When we look at the global distribution of topography and terrain types on Mars, further differences with the Earth and Venus emerge. There are essentially three major terrain units on Mars, distributed quite asymmetrically, as shown in Figure 5.47. Most of the southern hemisphere is occupied by an ancient terrain, apparently volcanic but subsequently heavily cratered by impacts (Figure 5.48), which (if the lunar cratering time-scale can be trusted here) date it at about 3 900 Ma. This terrain is responsible for the peak at around the median height on the hypsographic plot. Most of the northern hemisphere is occupied by lower-lying plains, partly filled by volcanic flows and partly by sediments (Figure 5.49), which are responsible for the peak at about 5 km below the median height on the hypsographic plot. These plains have a wide range of crater densities in different areas, indicating an age range mostly between about 3 800 Ma and 1 200 Ma.

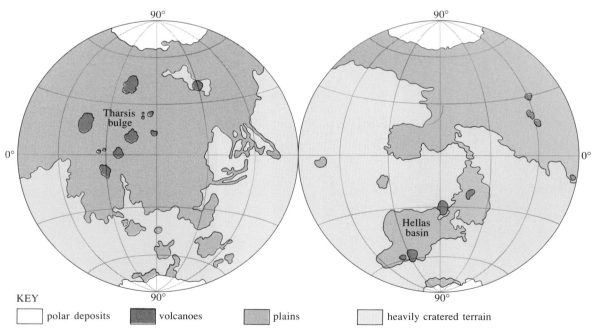

KEY
☐ polar deposits   ▨ volcanoes   ▨ plains   ☐ heavily cratered terrain

The division of Mars into markedly different regions — the high southern hemisphere and the low northern hemisphere — has become known as the **crustal dichotomy** (dichotomy means a division into two contrasting parts). The boundary between the two is very complex in detail, and in many places the ancient high terrain appears to be being eroded and shedding sediment into the lower plains. There are also a few large basins, up to 2 000 km across in the case of the **Hellas basin** (named on Figure 5.47), that appear to be plains material occupying ancient impact basins in a manner reminiscent of the lunar maria. Whether the origin of

Figure 5.47   The major terrain areas on Mars: heavily cratered terrain, plains, and volcanoes. Plains are covered by a mixture of lava and sediments. The Tharsis bulge and Hellas basin are physiographic features mentioned in the text.

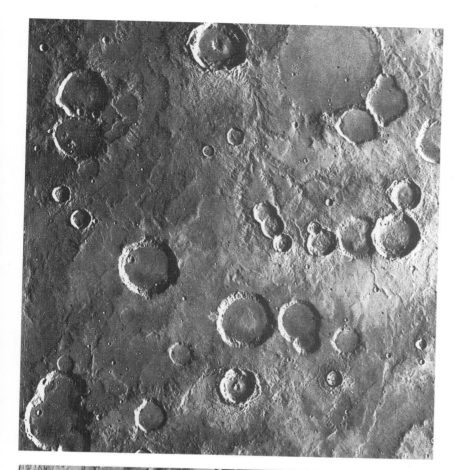

Figure 5.48 A *Viking Orbiter* image of part of the heavily cratered terrain on Mars, 250 km across.

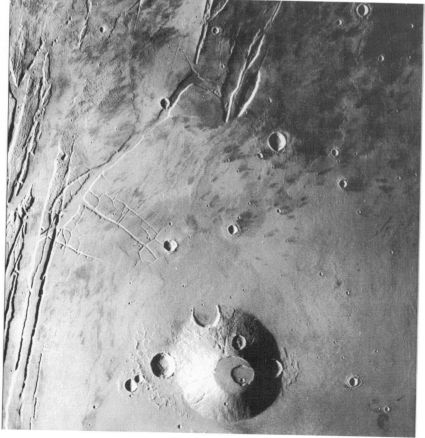

Figure 5.49 A *Viking Orbiter* image of part of the northern hemisphere plains area of Mars, 150 km across. This area has a lower crater density than the region shown in Figure 5.48, and includes an old volcano (with some impact craters superimposed) near the lower edge.

the crustal dichotomy itself can be explained by an enormous impact basin occupying the northern hemisphere, or whether it arises from an inherent division into two areas analogous to Earth-like continental-type crust (the southern hemisphere) and oceanic-type crust (the northern hemisphere), or has some other explanation, remains a matter of controversy. However, the degree of impact cratering and general lack of deformation make it clear that any plate-tectonic activity on Mars must have long since ceased.

The third of the three major terrain units on Mars consists of large volcanoes. Most of these occur on the plains unit of the northern hemisphere (e.g. the example in Figure 5.49), or around the boundaries of the Hellas basin. The youngest (to judge from the superimposed impact craters) are concentrated in a broad, 8 000-km-wide, 10-km-high, updomed region of the northern plains, known as the **Tharsis bulge**. This may owe its origin to updoming over a region of mantle upwelling, but whatever its origin, the Tharsis bulge is responsible for decreasing the proportion of the plains at low altitudes, which is why the low-altitude peak on the hypsographic plot (due to the plains) is so much less pronounced than the peak for the heavily cratered terrain. The volcanoes themselves account for the high-altitude 'tail' on this distribution. Four of the volcanoes on the Tharsis bulge reach about 29 km above the median height; the youngest of them, **Olympus Mons** (Plate 5.6), is on the north-western flank of the rise, and starts from a base altitude of less than 5 km. Thus, the drop from summit to base of Olympus Mons is some 24 km, well over twice the equivalent on Earth's tallest volcanoes, Mauna Kea and Mauna Loa, that form the main island of Hawaii, where the drop is about 9 km from their summits to the floor of the Pacific Ocean. Their relative dimensions are shown in Figure 5.50.

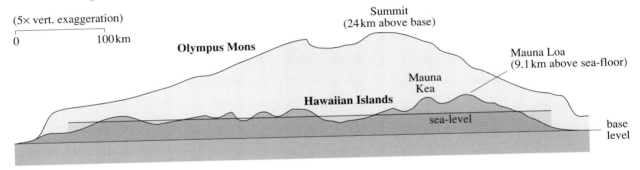

Figure 5.50 The relative sizes of Olympus Mons on Mars and the volcanoes forming the Hawaiian Islands.

All of the larger volcanoes on Mars have gentle slopes, indicating that they are formed mostly by low-viscosity lava flows, so they are generally regarded as basaltic shields of the kind exemplified by Hawaii on the Earth and various comparable examples on Venus (e.g. Plates 5.4 and 5.5). The density of superimposed impact craters shows that the construction of the martian volcanoes spanned most of the planet's history, ranging from about 3 500 Ma down to around 100 Ma in the case of the most recent eruptions on Olympus Mons. In fact, it is by no means obvious that Olympus Mons is extinct, and it may continue to erupt at intervals of a few tens or hundreds of millions of years indefinitely.

❑ The Tharsis bulge is generally reckoned to be situated over a site of deep mantle upwelling, that is to say a hot spot. The volcanoes on this rise span an age range of some 2 000 Ma, and they do not lie in any particular age-related pattern. What does this imply about any possible motion of the martian lithosphere over the deep mantle?

■ There cannot have been much motion over this time period. This is in marked contrast to the Hawaiian Ridge (and others that you met in Block 2, Section 2.2.2), where there is a consistent age sequence of volcanic islands and seamounts along the ridge demonstrating absolute plate motion at a rate of some 70 mm yr⁻¹ (which would amount to some 140 000 km over 2 000 Ma!).

5.9.2   THE THICKNESS AND HISTORY OF THE MARTIAN LITHOSPHERE

The fact that the martian volcanoes have been able to build up to such great heights can be partly explained by their remaining for so long above a source of magma (a long-lived mantle plume), as a result of the general immobility of the thick lithosphere. However, their height can also tell us the *depth* to the magma source, which, above a hot spot, is likely to be at the base of the lithosphere. To see how this works, you should now attempt ITQ 30.

ITQ 30

The reason why magma rises through the lithosphere is that it is less dense than the surrounding rocks. However, it can reach the surface only if there is sufficient pressure at its source, due to the weight of the lithosphere. Figure 5.51 explains this. The lithosphere is taken to have an *average* thickness t, and the magma source is taken to be immediately below the base of the lithosphere. If the average density of the rock in the lithosphere is ρ_r, then the pressure, P_r, at the base of the lithosphere is given by $P_r = t\rho_r g$, where g is the surface gravity on Mars. (Strictly speaking, the value of g we use should be intermediate between that at the surface and that at the base of the lithosphere, but this is a small effect, and to keep the calculation simple we will ignore it.)

In order for this pressure to be sufficient to force magma to the summit of a volcano of height s above the local average surface of the lithosphere, then this pressure must be great enough to overcome the pressure, P_m, at the base of a column of magma extending from the summit down to the magma source. The total height of this column of magma is $s + t$, so the pressure exerted by the magma column is given by $P_m = (s + t)\rho_m g$, where ρ_m is the density of the magma (ignoring any slight increase in magma density with depth).

Using $P_r = P_m$ as the limiting case when magma only just reaches the surface, calculate t, using $s = 24$ km (the height of Olympus Mons above its base), $\rho_r = 3.2 \times 10^3$ kg m⁻³ (an average value for mantle-plus-crust lithosphere) and $\rho_m = 2.8 \times 10^3$ kg m⁻³ (basalt). (You do not need to use a value for g.)

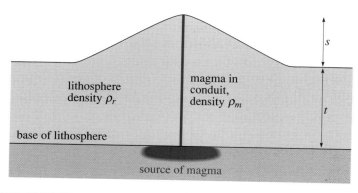

Figure 5.51   Lithostatic pressure, $t\rho_r g$, and magma pressure, $(s + t)\rho_m g$, under a volcano. For use with ITQ 30.

According to this calculation, the depth to the source of magma below Olympus Mons, and, by implication, the thickness of the lithosphere, is close to 170 km. If the average density of the rocks is less than the value used here, then the depth must be correspondingly greater. The apparently great thickness of Mars's lithosphere is confirmed by the gravity field data that show that neither the Tharsis bulge nor the youngest of the basins are isostatically compensated.

❑   The Hellas basin (about 3 800 Ma old) and the heavily cratered terrain (about 3 900 Ma old) *are* isostatically compensated. What does this suggest about the history of Mars's lithosphere, and how does it tally with the way in which you would expect Mars's heat flow to change over time?

■   Early in its history, Mars's lithosphere must have been thinner and/or less rigid than it subsequently became, allowing ancient topography to be brought into isostatic equilibrium. This is entirely compatible with a declining rate of radiogenic heat generation resulting in a progressively lesser rate of temperature increase with depth (i.e. a less steep geothermal gradient) over time.

SUMMARY OF SECTION 5.9

Mars is intermediate in size, mass and density between the Moon and the Earth, and may be formed of more highly oxidized material. Mars also shows a marked crustal dichotomy between high, ancient, heavily cratered terrain and generally lower-lying and younger plains areas. It has a wider range of topographic heights than either the Earth or Venus.

An updomed area within the plains area, the Tharsis bulge, is the site of the youngest volcanoes on Mars, among which Olympus Mons may not yet have ceased activity. Mars's lithosphere is thick and rigid, and shows no sign of being broken into plates or of major lateral movement relative to the deep mantle.

OBJECTIVES FOR SECTION 5.9

When you have completed this Section, you should be able to:

5.1   Recognize and use definitions and applications of each of the terms printed in the text in bold

5.18   Discuss the possible structure and composition of Mars, and cite some of the relevant evidence.

5.19   Discuss the possible history and thickness of Mars's lithosphere, citing the relevant evidence and chains of reasoning.

5.20   Make a qualitative comparison of the relative importances of conduction, plate recycling, and hot-spot volcanism in transferring heat from Mars's interior to the surface.

Apart from Objective 5.1, to which they all relate, the two ITQs in this Section test the Objectives as follows: ITQ 29, Objective 5.18; ITQ 30, Objective 5.19.

You should now do the following SAQs, which test other aspects of the Objectives.

SAQs for Section 5.9

SAQ 21 *(Objectives 5.1 and 5.18)*

What two independent lines of evidence can be used to argue that Mars as a whole is more oxidized than the Earth?

SAQ 22 *(Objectives 5.1 and 5.19)*

What are (a) the Hellas basin, and (b) the Tharsis bulge, and how may each have formed?

SAQ 23 *(Objective 5.19)*

What is the significance of the height of Olympus Mons in terms of the probable thickness of Mars's lithosphere?

SAQ 24 *(Objectives 5.1 and 5.19)*

(a)   Around the edges of the Tharsis bulge, all the discernible lava flows and water-cut channels, even the most ancient ones, indicate flow in downhill directions consistent with the present topography. What does this indicate about the age of the Tharsis bulge?

(b)   How can this age be reconciled with the lack of complete isostatic compensation shown by the Tharsis bulge?

SAQ 25 *(Objectives 5.19 and 5.20)*

(a)   Rank the three methods of outward heat transfer from Mars's interior (conduction, plate recycling, and hot-spot volcanism) in decreasing order of importance as heat loss mechanisms on Mars today.

(b)   Discuss how this compares with the probable situation 3 800 Ma ago.

5.10 Io

Io, usually pronounced 'eye-oh', is the innermost large satellite of Jupiter. Most of our knowledge about this body and about the other satellites of the outer planets comes from two space probes, *Voyager 1* and *Voyager 2*, that were launched in 1977 (Figure 5.52). They both visited the Jupiter (1979) and Saturn (1980, 1981) systems, and *Voyager 2* was able to continue on to the Uranus (1986) and Neptune (1989) systems as well.

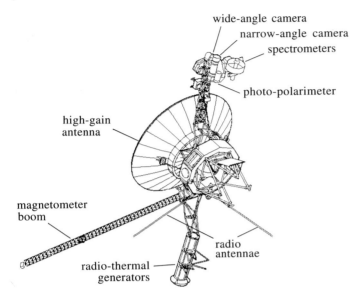

Figure 5.52 One of the two *Voyager* spacecraft. Each was about the size of a minibus, and had a mass of 825 kg.

5.10.1 DATA ON OUTER PLANET SATELLITES ACQUIRED BY THE *VOYAGER* MISSIONS

The *Voyager* probes were equipped with a variety of instruments. Their electronic cameras provided the most spectacular results, by way of images of the surfaces of the planets and their satellites (as you saw in AV 10 and VB 07), and other important data were provided by magnetometers and charged-particle detectors (which investigated the electric and magnetic conditions of interplanetary space), thermal infrared detectors (which measured temperatures) and radio-tracking instruments (which enabled the trajectories of the space probes to be monitored precisely).

The trajectories of the *Voyager* probes were chosen so as to pass by satellites in succession (Figure 5.53), but the probes were travelling fast,

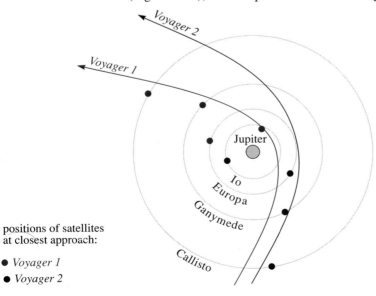

positions of satellites at closest approach:

● *Voyager 1*
● *Voyager 2*

Figure 5.53 The trajectories of *Voyager 1* and *Voyager 2* through the Jupiter system, showing the positions of the four major satellites at the time of each encounter.

in excess of 50 000 km h^{-1}, and so were able to get only fleeting glimpses of some of them. The images and other data obtained during these missions confirmed that the major satellites are all in synchronous rotation, and generally gave us rather good measurements of their sizes. The slight perturbations caused by the gravitational influence of each satellite on the spacecraft trajectories provided somewhat poorer estimates of their masses, so that the resulting uncertainties in their densities are often as much as 20% or more. Moreover, the gravity data are inadequate to provide any indication of the moment of inertia or internal density distribution of any of the satellites, so our present models for their internal structures are based simply on their bulk density and assumed composition.

5.10.2   THE NATURE OF IO

As you discovered in answering ITQ 12, Io has a size and density that rank it with the terrestrial planets. It has an extremely tenuous atmosphere, mainly of sulphur dioxide, with a surface pressure of 10^{-2} N m^{-2} (i.e. 10^{-7} of the Earth's atmospheric pressure), which is insufficient to shield the surface from cosmic radiation and charged particles that are funnelled towards Io by Jupiter's magnetic field. Unlike the other major satellites of the outer planets, Io's reflectance spectrum shows no sign of water; instead, its colour is distinctly reddish, with an absorption feature in the ultraviolet that indicates the presence of sulphur.

Io's density suggests that it is a largely silicate body, and is great enough to allow the possibility that it has a small core of Fe or FeS.

❑   Comparing the densities of Io and the Moon in Table 5.2, and assuming that they each have a core of similar composition, which body is likely to have the largest core, relative to its total size?

■   As Io has the higher density, its core should be relatively larger than the Moon's.

Io is fractionally larger than the Moon, but the greater relative size of Io's core probably means that the total mass of silicates within Io is similar to that of the Moon.

ITQ 31

(a)   Assuming that the silicates in Io and the Moon have similar concentrations of heat-producing elements, what does this suggest about the relative rates of global radioactive heat production in these two bodies at the present day?

(b)   Would you therefore expect Io to be volcanically active?

One of the more remarkable discoveries made by the *Voyager* space probes was that Io does in fact have several active volcanoes. There were eruption plumes over nine volcanoes when *Voyager 1* flew by in March 1979 (Figure 5.54, Plates 5.7 and 5.8), eight of which were still active four months later during the *Voyager 2* encounter. Analysis of the way in which sunlight is scattered within these plumes suggests they are dominated by particles up to 1 μm in diameter. The shapes of the plumes suggest that these particles are blasted upwards in an explosive gas jet and then fall out on ballistic trajectories.

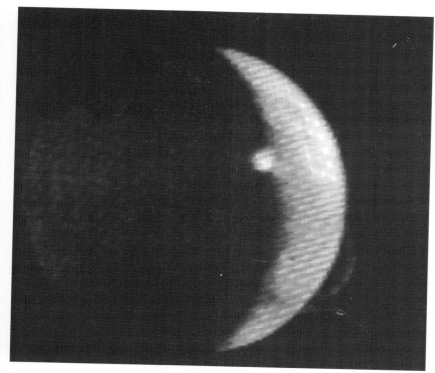

Figure 5.54   Two umbrella-shaped plumes being erupted from volcanoes on Io, seen by *Voyager 1* in March 1979. One is the bright spot just within the shadowed area of the globe, and the other is seen faintly in outline beyond the lower right edge of the disc.

Figure 5.55   An infrared image of Io, recorded using a telescope in Hawaii in 1989. The bright spot on the image corresponds to the position of a volcano shown to be active during the *Voyager* fly-bys.

The *Voyager* probes provided further evidence for active volcanism by mapping the surface temperature on Io, using thermal infrared detectors. These had low resolution, averaging out the radiation across a measuring spot 70 km or more across, but nevertheless demonstrated temperatures in excess of 600 K at the sources of some of the plumes and detected several other 'hot spots' that were not associated with plumes. These 'hot spots' stand in marked contrast to the rest of Io's surface, which has a daytime temperature of about only 120 K (−153 °C). Subsequent to the *Voyager* encounters, Earth-based infrared telescope techniques have improved sufficiently to demonstrate that several 'hot spots' continue to be active today (Figure 5.55).

As you saw in the VB 07, 'Other Worlds', there is a generally accepted explanation of how enough heat is generated within Io (by a non-radioactive process) to cause all the volcanism, and we shall examine this in more detail towards the end of this Section. Before that, we shall look at the effects of the volcanic processes. Eruption plumes and hot spots are not the only clues that Io is a geologically active world. Prior to the *Voyager* encounters, ground-based observations had detected a cloud of sodium atoms surrounding Io, and a doughnut-shaped belt ('torus') of ions (notably ionized sulphur and oxygen) spread out along Io's orbit (Figure 5.56). These atoms and ions are now thought to be supplied by eruption plumes of the kind shown in Figure 5.54 and Plates 5.7 and 5.8, a process that is compatible with the probable presence of sulphur dioxide 'frost' on parts of Io's surface as revealed by examination of Io's reflectance spectrum, which differs markedly from the ice-

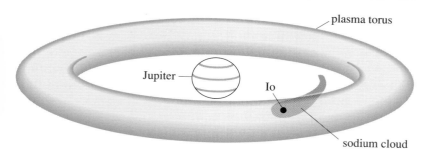

Figure 5.56   The sodium cloud and plasma torus (a belt of ionized atoms) associated with Io. The ions forming the plasma torus are held there by Jupiter's strong magnetic field, which is also responsible for giving them their charge in the first place.

dominated spectra of the other satellites of the outer planets (e.g. Figure 5.25). There is a further line of evidence that Io is a geologically active world, which you should be able to deduce by attempting the next ITQ.

ITQ 32

Examine the images reproduced in Figure 5.57 and Plates 5.7 and 5.8. What do these tell you about the nature and probable age of Io's surface?

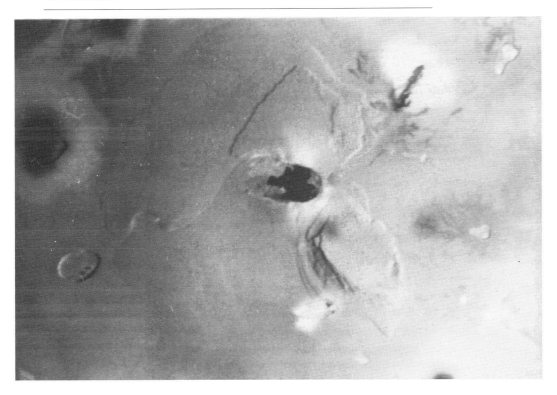

Figure 5.57 *Voyager* image showing part of the surface of Io, about 700 km across. For use with ITQ 32.

Thus, Io's surface appears to be dominated by the products of volcanic eruptions, both lava flows and particles dispersed by the eruption plumes. Such craters as there are have none of the characteristics of impact structures; they are less regular in shape, many appear to have been the sources of lava flows, and most of the 'hot spots' detected by the *Voyager* instruments are located within them. Any crater like this on Io is usually regarded as a volcanic **caldera**, that is to say a crater formed by collapse due to withdrawal of underlying magma or as a consequence of an explosive eruption (many calderas are known on Earth). Volcanic activity on Io is evidently sufficient to have obliterated all traces of impact craters, and it has been calculated that the current deposition rate of new material over Io's surface must be at an average rate of at least 1 mm yr^{-1} to account for this.

❑ Do you think that volcanic activity is the only process that could account for the obliteration of impact craters on Io?

■ In the near-total absence of an atmosphere and associated erosion and deposition processes, only tectonic activity of some kind could be an additional contributor to the destruction of impact craters.

In fact, no clear signs of major crustal deformation on Io are revealed on any of the images. Traces of tectonism are rare, being limited to a few surface breaks that probably represent small faults. Certainly, an organized pattern of plate tectonics is absent; there are no spreading axes, subduction zones or fold- and thrust-belts, and the volcanoes are dispersed widely across the surface.

5.10.3 VOLCANIC PROCESSES ON IO

But what of Io's volcanic processes? Here we reach one of the major controversies in planetary geology, which involves two competing schools. One holds that the lava flows on Io's surface are composed of sulphur rather than silicates, citing as evidence the reddish colour of Io's surface, the sulphur in the plasma torus, and the temperatures recorded by infrared techniques at 'hot spots' that are generally well below the solidus temperatures of silicates (on Earth, flows produced by molten sulphur are extremely rare, but not unknown). The opposing school claims that Io's flows are more conventional lava, i.e. some kind of silicate, pointing to the steep inner slopes of many of the calderas that could not reasonably be composed of such a weak material as sulphur, and appealing to crusting over of much of the surface of molten lava in 'hot spots' to lower the apparent temperature (as indeed happens in volcanic lava lakes on the Earth).

It is possible that the eruption plumes provide the key to resolving this controversy. Volcanic eruption plumes on the Earth are driven by the explosive escape of volcanic gases (dominated by water vapour) when magma comes close to the surface. In the case of Io, the apparent ubiquity of sulphur and sulphur dioxide on the surface and surrounding it in space indicates that these, rather than water, are the most important volatile phases. It has become recognized that there are two varieties of eruption plume on Io, a long-lived comparatively low-temperature kind reaching heights of 50–120 km and a short-lived higher-temperature kind reaching 300 km in height like the one shown above Pele in Plates 5.7 and 5.8. It has been suggested that the low-temperature variety of plume is driven by the escape of sulphur dioxide heated to boiling point by the intrusion of molten sulphur (melting point 393 K). Liquid sulphur dioxide, which exists between 198 K and 263 K when subjected to a pressure of 10^5 N m^{-2}, has a very low viscosity (about half that of water) and so could move freely through the substrate, being drawn in from a wide area over a long time period to replenish continuously the supply fed into the plume. On the other hand, the short-lived and higher-temperature Pele-type plumes are more likely to be driven by vaporized sulphur. This is likely to be a result of the intrusion of molten silicate magma. The sulphur in contact with this magma must have been heated to above 700 K, otherwise the resulting vapour would not be able to fragment the overlying liquid sulphur, and the only thing erupted would be a frothy flow of liquid sulphur. Because liquid sulphur has a considerably higher viscosity than sulphur dioxide, it could not move so freely through any substrate, so a Pele-type plume would be likely to deplete the reservoir of sulphur within a matter of weeks, thus explaining the comparatively short duration of Pele-type eruptions.

Therefore, the (admittedly limited) evidence from the eruption plumes suggests that both sulphur and silicates occur as magmas at shallow depths within Io's crust.

☐ If this is correct, what does it suggest about the nature of the lava flows seen on Io's surface?

It seems reasonable that some of them could be sulphur and others could be silicates.

This matter is unlikely to be resolved for some time. High-resolution infrared imaging of a flow actually being emplaced would help to constrain the temperature and hence the composition. Better spectral observations of the surface would be unlikely to give any convincing proof of the nature of a flow, because even a silicate lava would be likely to become coated by sulphur and sulphur compounds under the remarkably hostile environment of Io's surface.

5.10.4   THE HEATING MECHANISM FOR IO

By now, it should be clear to you that Io is a rather remarkable world. For its size, it is far more active volcanically than the Earth. The global average of Io's heat flow has been estimated at $1-3\,W\,m^{-2}$, from the volcanoes alone.

ITQ 33

How does this compare with the value for the present-day average heat flow for the Moon that you worked with in Section 5.7.1?

Thus, Io's heat flow is about two orders of magnitude greater than the Moon's. Clearly, it would be unreasonable to expect this to be due to radiogenic heating (which probably contributes only about 1% to Io's total observed heat flow), and unless Io is actually a young world that accreted only recently (which would be very difficult to account for!) we can rule out accretional heat or heat from core formation and the like as viable explanations for Io's heat flow. The accepted answer lies in a special form of an alternative method of heat production that you met in Section 5.7 and in VB 07, 'Other Worlds'.

❑   Can you recall what this mechanism is?

■   The answer lies in the dissipation of tidal energy

In the case of the Earth–Moon system, tidal energy is dissipated at a low rate by gradual slowing down of the Earth's rotation associated with a gradual increase in the radius of the Moon's orbit. However, the situation in Jupiter's satellite system allows tidal energy to be dissipated at a much greater rate.

ITQ 34

Compare the orbital periods of the three inner satellites of Jupiter, as given in Table 5.2 (p. 41). Can you see any mathematical relationship between these?

The answer to this ITQ shows that the inner satellites of Jupiter exhibit a phenomenon known as **orbital resonance**. As a result, every time Io passes close to Europa it does so in the same part of its orbit (Figure 5.58). When this happens, Io experiences a slightly increased gravitational tug from Europa, the effect of which is to distort Io's orbit into an ellipse. In the same way, Europa's orbit is made elliptical by its regular interaction with Ganymede. This **forced eccentricity** is a very important

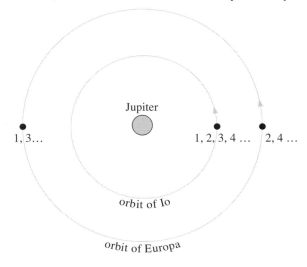

Figure 5.58   Orbital resonance between Io and Europa: Io passes Europa at virtually the same part of its orbit every time, and this repeated gravitational interaction distorts Io's orbit into an ellipse. The numbers show the positions of Io and Europa after Io has completed successive whole numbers of orbits. Io's orbit has a very low 'eccentricity' of only 0.0041, which means its shape departs from a circle by about only 0.4%, and so it appears circular when drawn without exaggeration, as here.

factor, because, in the absence of such a perturbation, their orbits would long since have become effectively circular (as a result of tidal interactions with Jupiter), and the present rate of dissipation of tidal energy would be extremely low.

ITQ 35

Conservation of angular momentum dictates that a body such as Io must spin on its axis at an effectively constant rate. However, the laws governing a satellite's orbital motion mean that its speed is greatest when it is closest to its planet and least when it is furthest from its planet. Can you see what effect these two conditions must have on the supposedly synchronous rotation of the satellite?

This wobble, or **libration**, of a synchronously rotating satellite such as Io leads to tidal heating. The strong gravitational pull of Jupiter raises a tidal bulge of probably a few kilometres in height on Io's Jupiter-facing hemisphere (and an equivalent one in the opposite hemisphere), and it is the tidal drag on this bulge that brought Io into synchronous rotation in the first place.

ITQ 36

What effect will libration have on the position of the tidal bulges, if they are to remain in equilibrium?

Thus, the bulges will attempt to change their position on Io's surface, by distorting Io's shape. An additional consideration is that, because the tidal force will be greatest when Io is closest to Jupiter, the height of the bulges will vary according to Io's position along its elliptical orbit. Both of these phenomena involve continual changes in Io's shape, and therefore, by tidal dissipation, work is being done all the time to deform Io's interior.

The dissipation of all this energy results in a continuing input of heat to Io, and the whole process is often referred to simply as **tidal heating**. The rate of energy dissipation depends inversely on Io's rigidity and on various other properties, the values of which are poorly constrained. The equally poorly understood internal properties of Jupiter are also significant factors. An early model (which was published just days before *Voyager 1* discovered Io's eruption plumes) suggested that tidal heating would make Io's interior entirely molten below a lithosphere only 10 km or so in thickness. The presence of mountains up to 10 km high on Io (e.g. top right of Plate 5.7) seems to rule this out, if the mountains are to be even partially isostatically compensated. It appears more likely now that Io has a lithosphere about 50 km thick and that much of the tidal energy is dissipated below it.

The possible internal structure of Io is shown in Figure 5.59. Undoubtedly, the degree of heat production has been sufficient to allow Io to evolve into an internally differentiated body. The core may be molten. The lower mantle is presumed to be solid and relatively rigid, and much of the tidal dissipation may occur here. Above it, the asthenosphere must be convecting rapidly (to transport the heat outwards), and there may even be a thin liquid shell. Flexing between the overlying lithosphere and the asthenosphere provides another site for dissipation of tidal energy.

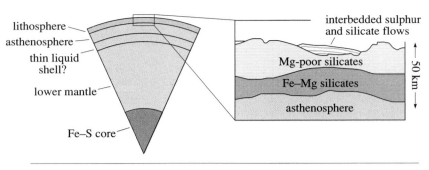

Figure 5.59   A model for Io's structure. We have no seismic or moment of inertia data for Io, so this is constrained only by Io's density. Because of its demonstrably high heat flow, Io is assumed to be fully differentiated. In the view on the left, the thickness of the layers above the lower mantle has been exaggerated for clarity, but the scale is uniform in the inset on the right.

ITQ 37

How does the thickness of Io's lithosphere as implied by Figure 5.59 compare with the thickness of the lithospheres of the Earth and the other terrestrial planets you have investigated in this Block?

We are not going to examine Mercury in this Course, but we can point out here that its thermal lithosphere is probably about 120–170 km thick. Io's lithosphere (*according to this model*) is therefore the thinnest of all the terrestrial planets, with the possible exception of the thermal lithosphere of Venus. In view of this, it is perhaps surprising that Io shows so few indications of tectonic processes of the kind that might be expected in a thin lithosphere above a presumably vigorously convecting asthenosphere.

5.10.5   SULPHUR AND IO'S CRUST

As you have read, the extent of sulphur in Io's crust is a matter of some dispute. If Io's composition is approximately chondritic (but without water; see Section 5.6), then its sulphur content should be about 5% by mass. If gathered together, this would be sufficient to make a 50 km thick layer of sulphur over the surface.

❑ Do you think it likely that such a thickness of sulphur could exist as part of Io's crust, and can you suggest where else sulphur might be concentrated in Io?

■ It is unlikely. Much of the sulphur is probably trapped in the core, as may occur in the Earth (Block 1, Section 1.11.3) and Mars (Section 5.9). In addition, some of Io's sulphur is evidently being lost to the plasma torus (Section 5.10.2), but it is unclear whether this amounts to a significant long-term loss.

The model for Io's structure in Figure 5.59 assumes that the bulk of Io's crust is composed of silicates, though it does not rule out the possibility that sulphur flows are an important feature. It has been calculated that conduction through any feasible lithospheric material would be inadequate to transport the heat generated within Io towards the surface at the observed rate without resulting in progressive melting upwards from the base of the lithosphere, until it became much thinner than its likely present thickness.

❑ Can you suggest the most reasonable way for the bulk of Io's heat to be transported towards the surface through the lithosphere?

■ The answer lies in magmas originating near the lithosphere–asthenosphere boundary rising up towards the volcanoes (i.e. hot-spot volcanism, possibly above mantle plumes), and indeed the high value assumed for Io's high heat flow is actually based on measurements of the heat output at volcanoes (Section 5.10.4).

Quite apart from the observed volcanic processes, logic suggests that Io has a geochemically distinct crust occupying the upper part of its lithosphere. This is indicated as Mg-poor silicates in Figure 5.59, and the Fe–Mg silicate layer below it is effectively the lithospheric part of the upper mantle.

5.10.6 CLOSING REMARKS

We have discussed Io at comparatively great length because it is more active volcanically than the Earth or any other terrestrial planet. We hope that this Section does not leave you with the impression that Io is well understood; there are many problems outstanding. Among these are that the currently favoured models for the present rate of tidal energy dissipation within Io suggest that it is about an order of magnitude *less* than the observed heat output. This could be due to incorrect assumptions about the internal structures of Io and Jupiter, or it could indicate that Io's heat output varies significantly over time and is currently anomalously high (after all, we have been able to observe it only since the mid-1970s when ground-based infrared techniques became adequate), in which case it may be subject to episodic pulses of volcanic activity. Just to indicate some of the further complexities involved: if Io's heat output does occur in pulses, then this could either be a result of changes in the convective state of its asthenosphere or reflect changes in the rate of tidal energy input, which would result from evolution of the orbital interactions between Io and Europa (and between Europa and Ganymede). With several bodies involved, about whose internal properties we can do little more than speculate, the system is far too intricate to model in detail, but it does seem that Io's orbit is liable to undergo periods of comparatively high eccentricity and therefore higher rate of heat production (as at present) of about 20 million years separated by intervals of low eccentricity of perhaps 100 million years.

SUMMARY OF SECTION 5.10

Io is volcanically active, as demonstrated by observations of eruption plumes and localized high temperatures. Volcanic resurfacing has obliterated all traces of impact craters on Io, because it occurs at a rate sufficient to keep pace with the average rate of production of impact craters. The eruption plumes are probably driven by escaping sulphur dioxide or sulphur vapour, but it is not clear whether 'lava' flows seen on the surface are made of silicates or sulphur. The steepness of the inner walls of volcanic calderas and the height of isolated mountains suggests that Io's crust (and the rest of the lithosphere) is constructed largely of silicates, rather than sulphur (which would be too weak to produce such steep landforms).

Io's heat flow is about two orders of magnitude greater than can be explained by radiogenic heat production in a body of this size, and is attributed to the dissipation of tidal energy within Io. Io's tidal energy is a consequence of orbital resonance between Io and Europa, which prevents Io's orbit becoming circular. The tidal bulges raised on Io by Jupiter's gravity are continually being deformed due to the mismatch between Io's constant rate of captured rotation and the varying speed of Io's orbital motion. Over tens of millions of years, changes in orbital configurations may lead to significant variations in the rate of heating by tidal dissipation. Despite Io's volcanic activity, it shows little evidence of tectonic processes.

OBJECTIVES FOR SECTION 5.10

When you have completed this Section, you should be able to:

5.1 Recognize and use definitions and applications of each of the terms printed in the text in bold.

5.21 Appreciate the causes and possible contributions of various heating mechanisms to Io's heat output.

5.22 Describe a plausible model for Io's internal structure, and cite relevant evidence.

5.23 Discuss the likely relative importances of silicates and sulphur in Io's volcanic processes.

5.24 Compare the relative importances of conduction, plate recycling, and hot-spot volcanism in transferring heat from Io's interior to its surface.

Apart from Objective 5.1, to which they all relate, the seven ITQs in this Section test the Objectives as follows: ITQs 31 and 33–36, Objective 5.21; ITQ 37, Objective 5.22; ITQ 32, Objectives 5.23 and 5.24.

You should now do the following SAQs, which test other aspects of the Objectives. ·

SAQS FOR SECTION 5.10

SAQ 26 (*Objective 5.21*)

Io's total power output, as measured by infrared techniques, is at least 4×10^{13} W. The total accretional heating feasible for Io's formation is estimated to be about 7×10^{28} J.

(a) If all the accretional heat were released at a uniform rate over the age of the Solar System (1.4×10^{17} s), what would be Io's average rate of loss of accretional heat over this time? (*Note:* $1 \text{ W} = 1 \text{ J s}^{-1}$)

(b) How does this compare with measured Io's present rate of heat loss?

(c) Would you expect accretional heat to be released at a uniform rate, and how does this consideration affect the comparison between the contributions to present-day heat flow between accretional heating and tidal dissipation?

SAQ 27 (*Objective 5.23*)

How does the contention that most of Io's internally generated heat is carried toward the surface by the upwards movement of silicate magmas tally with the proposed causes of the eruption plumes discussed in Section 5.10.3?

SAQ 28 (*Objective 5.22*)

Taking into account the current rate of resurfacing of at least 1 mm yr^{-1}, and other factors, explain whether you would classify Io's crust as a primary, secondary or tertiary crust.

SAQ 29 (*Objective 5.24*)

Figure 5.60 is a triangular plot on which you are asked to indicate the relative importance of conduction through the lithosphere, plate recycling and hot-spot volcanism as methods of heat loss in several planetary bodies. A point for the Earth has already been drawn on the Figure, indicating that most of the Earth's heat loss occurs by means of plate recycling, with a smaller contribution from conduction and only a tiny fraction by way of hot-spot volcanism. Now add points representing the relative contributions of these three heat loss mechanisms for Io, the Moon, Venus and Mars.

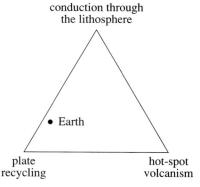

Figure 5.60 The relative importance of the three heat loss mechanisms in planetary bodies. To be completed in SAQ 29.

5.11   ICY SATELLITES

There are 17 icy satellites in the outer Solar System in excess of about 200 km in radius, and the class expands to 18 if we include Pluto (which, as remarked in Section 5.6, seems to be similar to Neptune's satellite Triton). Obviously, there is not sufficient space here to deal with these in any detail. Instead, we will take a brief look at just a few, treating them as examples of the variety of ways in which planetary bodies with recognizable lithospheres and asthenospheres can respond to internal heating.

5.11.1   VOLCANISM ON ICY WORLDS

At the end of Section 5.6, you were introduced to the idea that in the outer Solar System the temperature is so low that the outermost ice in an icy satellite will behave as a lithosphere, and that deeper ice can flow convectively like an asthenosphere if the internal temperature is sufficiently high. This means we can expect to see evidence of a wide range of tectonic activity, notably faults of various kinds. But what about volcanism? What is the icy equivalent of lava?

The obvious example is water. In theory, water could be erupted at the surface of an icy satellite either in an explosive fashion (driven by gases), to produce the equivalent of a terrestrial pyroclastic eruption in which the surface would be showered by icy debris, or as an effusive flow that travelled some distance before it solidified, thereby creating a fresh icy 'lava' surface. There is actually a considerable variety of flow types possible, because the erupted product does not have to be 100% liquid. It could contain any proportion of ice crystals, ranging up to nearly 100% crystals lubricated by a tiny fraction of liquid water (or other fluid), in which case the 'lava' would be orders of magnitude more viscous than water and would spread much less far.

Another important factor is the inevitable presence of impurities within the ice. For example, the rocky material with the icy satellites, if initially of generally chondritic composition, will have contained abundant water-soluble substances. Provided that the chemicals in the rocky component have been able to react with the H_2O in the ice (and heating during and subsequent to accretion probably gave ample opportunity for this), the present ice in an icy satellite is likely to have small salt crystals distributed within it. You have already met an ice–salt mixture, in Block 3, Section 3.4.2 (Figure 3.28), where you were introduced to the $NaCl–H_2O$ system. However, experiments with chondritic meteorites show that sodium chloride is likely to be no more than an extremely minor component in any solution derived by water interactions with chondritic rock. Instead, the salt solution, or **brine**, will be dominated by sulphates, especially of magnesium. Thus, in order to get an idea of the effects of dissolved salts in the ice within the satellites of the outer planets, we must consider the $MgSO_4–H_2O$ system. This is plotted in Figure 5.61.

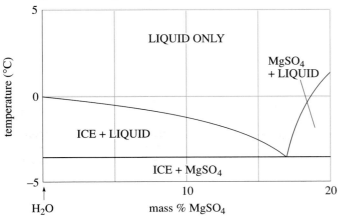

Figure 5.61   A portion of the $MgSO_4–H_2O$ system.

ITQ 38

With reference to Figure 5.61, at what temperature would frozen water containing crystals of $MgSO_4$ begin to melt, and what would be the composition of the first liquid to form?

In answering this ITQ, you should have realized that any melts forming within an icy satellite are unlikely to be pure water. The presence of salts within ice results in melts that are actually quite strong brines, and these can differ significantly from water in physical properties such as density and viscosity. The melt in ITQ 38 will be about six times more viscous than water, so it will flow about six times less freely. In the case of brines, the temperature at which the melt forms is only a few degrees below the melting point of pure water, but you will see in Section 5.11.4 that beyond Jupiter the presence of volatile impurities such as ammonia can lower the temperature at which melting will begin very considerably, thereby making it much easier for eruptions to occur. Thus, icy volcanism is similar in all its important characteristics to the silicate volcanism with which we are familiar on Earth, except that it occurs at much lower temperatures. To avoid confusion with terrestrial-style volcanic processes, icy volcanism is often referred to as **cryovolcanism** (using the Greek word *kryos*, meaning frost).

Cryovolcanism can occur on an icy planetary body provided two conditions are satisfied. The first is that there must be a potential 'magma' source close enough to the surface. This requires sufficient heat flow to drive a convecting asthenosphere. The second is that any melt generated must have access to the surface. This is helped by fracturing of the lithosphere, especially by extensional tectonics. The melt must also be buoyant.

❑   Water near its melting point is denser than pure ice by some 8%, and brines are about 10–15% denser still. Can you suggest how a melt could become buoyant relative to the near-surface material in an icy satellite?

■   There are two possibilities. The liquid could contain gas bubbles (produced by the escape of volatiles as the confining pressure decreases), or the ice could be weighed down by denser impurities. The former would lower the density of the liquid, the latter would increase the overall density of the ice. Although salt crystals within the ice will have some effect in increasing its overall density, dispersed rock fragments are probably even more important, especially in satellites whose outer region has not become fully differentiated (e.g. Figure 5.26a and c).

There are several examples of satellites showing the effects of cryovolcanism in the remainder of this Section.

5.11.2   EUROPA

As you may recall from SAQ 10, Europa represents something of a transition in composition between the terrestrial planets and the icy moons. It is worth examining in view of the rather striking characteristics of its lithosphere. Images of Europa recorded by the two *Voyager* probes revealed an icy surface with very subdued topographic relief and less than a dozen impact craters (all between 5 km and 30 km across), but cut by many dark and bright bands that probably represent fractures of some kind (Figure 5.62).

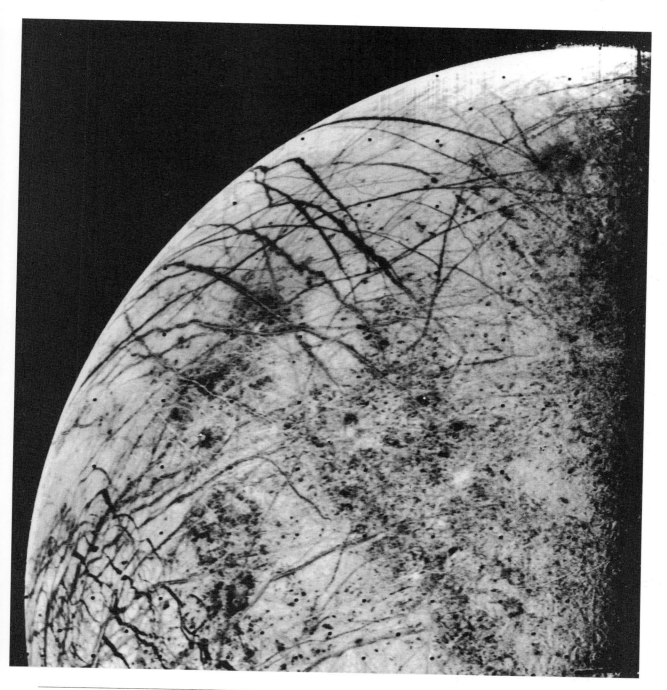

Figure 5.62 *Voyager 2* image showing part of Europa nearly 3 000 km across.

ITQ 39

(a)   At the low surface temperature prevailing on Europa (about 130 K), near-surface ice would behave rigidly, like rock over geological time-scales. What do the following tell you about the age of Europa's surface and the strength and thickness of its lithosphere?

(i)   The very low crater density.

(ii)   The very low relief.

(b)   What do these imply about the degree to which Europa has been geologically active over recent times?

Europa stands in marked contrast to the other somewhat larger icy satellites of Jupiter, Ganymede and Callisto, whose surfaces have high crater densities and show no trace of tectonic or volcanic activity over at least the past 2 000 Ma (Figure 5.63). The thinness of Europa's lithosphere, its apparently young surface and the presumably tectonic fractures criss-crossing its surface are all symptomatic of an anomalously high heat flow.

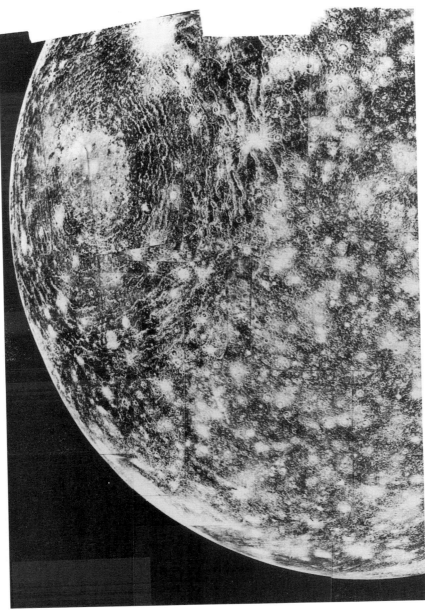

Figure 5.63   A mosaic of *Voyager 2* images showing much of Callisto. There is a large ancient impact basin towards the top left (surrounded by concentric fractures), and the whole surface is densely cratered.

❏   Look at the orbital periods of Europa and its neighbours in Table 5.2. Do these indicate a probable cause of internal heat generation for Europa?

■   Europa's orbital period is almost exactly half that of Ganymede and twice that of Io. This orbital resonance results in forced eccentricity of Europa's orbit, and heat is generated by the ensuing tidal dissipation (similar to the manner discussed for Io in Section 5.10.4).

A model for Europa's structure, consistent with its icy surface but high density, is shown in Figure 5.64. Unique among the icy satellites today, Europa is likely to have an asthenosphere of liquid water or brine (which, if it is liquid, must have a sharp upper boundary, unlike that of the Earth's asthenosphere). This is kept molten by heat from tidal dissipation in the rigid layers above and below, supplemented by radiogenic heat from Europa's large silicate lower mantle/core. As there are apparently no major hot-spot-like volcanic centres on Europa, it seems that transfer of heat outwards through the lithosphere is probably accomplished by a mixture of conduction and plate recycling. This is not to say that Europa shows Earth-style plate tectonics. There is abundant evidence of creation of new lithosphere (by cryovolcanic eruption) along the 'cracks' visible in images such as that in Figure 5.62, but little sign of areas where old lithosphere could be drawn back down into the asthenosphere. As in the case of Venus (Section 5.8.3), the buoyancy of Europa's icy lithosphere (of whatever age) probably inhibits the initiation of Earth-style subduction zones, in contrast to the Earth where (as you saw in Block 2, Sections 2.6.2 and 2.6.3) the cooling of the lithosphere with age eventually eliminates its buoyancy and leads to the dominance of the slab-pull force. However, less than half of Europa was imaged at adequate resolution by the *Voyager* probes, and it is possible that Europa's equivalent of subduction zones remain to be discovered by a future mission.

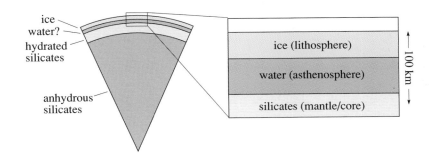

Figure 5.64   A model for the internal structure of Europa.

5.11.3   MIMAS AND ENCELADUS

We now move to the opposite end of the size spectrum, and compare two of the smallest near-spherical icy satellites, Mimas (197 km radius) and Enceladus (251 km radius). These are the innermost sizeable satellites of Saturn, and each has a low density, $1.2 \times 10^3 \, \text{kg m}^{-3}$, which, as you saw in ITQ 13, suggests they are composed of an approximately 7 : 1 ice : rock mixture by volume. Because they are so small and contain so little rock, radiogenic heating would be totally incapable of maintaining the interior ice at a sufficient temperature to convect, except possibly in their first few million years. Their orbital periods are listed in Table 5.2 (p. 41).

❑   Can you see any orbital resonances among Saturn's inner satellites?

■   To a close approximation, Mimas is in 2 : 1 orbital resonance with Tethys, and Enceladus is in 2 : 1 orbital resonance with Dione.

This was well known before the *Voyager* encounters with the Saturn system, but was not thought likely to be a significant cause of tidal heating in either Mimas or Enceladus because in both cases the degree of forced eccentricity of their orbits is very small. Thus, with radiogenic heating insignificant and tidal dissipation apparently too slight, both these satellites were expected to be densely cratered worlds with no signs of internally driven activity. This prediction turned out to be correct in the case of Mimas as you can see in Figure 5.65, but was wide of the mark for Enceladus as Figure 5.66 demonstrates.

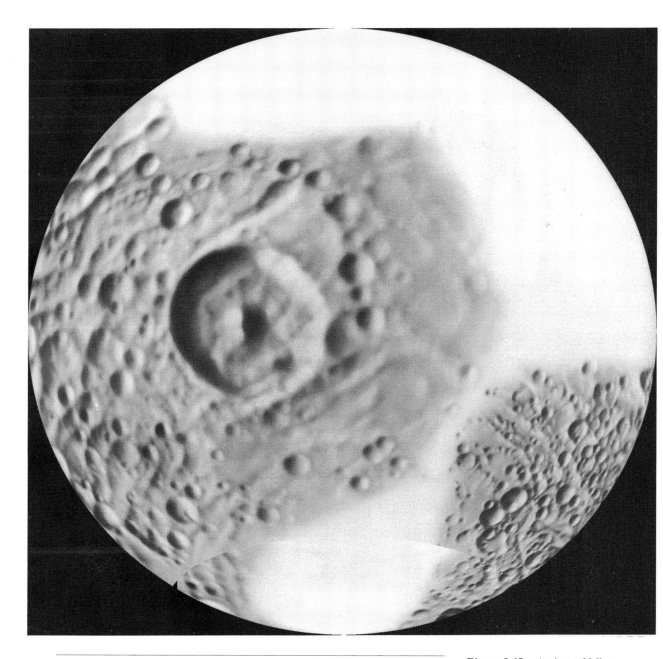

ITQ 40

It should be apparent from Figure 5.66 that Enceladus has an interesting geological history. This image has been interpreted for you in Figure 5.67. Examine these two Figures and describe the evidence for the *time-span* and *nature* of activity, paying particular attention to the distribution and shapes of craters, and the tectonic features and cross-cutting relationships that you can see. You should attempt, so far as is possible, to rank the terrain units identified in Figure 5.67 in order of age.

Figure 5.65 A view of Mimas, showing a surface totally dominated by impact craters. The very large crater near the centre is 130 km in diameter. It has been calculated that an impact slightly larger than the one that created this crater would have had sufficient energy to break Mimas into fragments. There are no data for the blank areas.

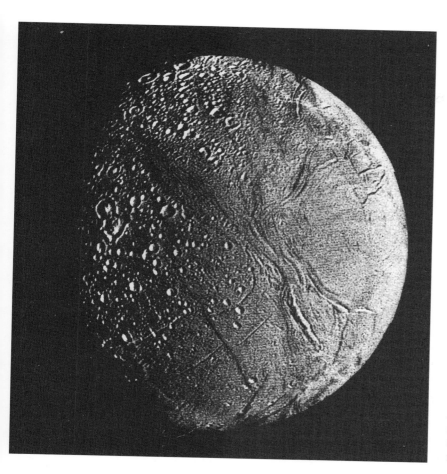

Figure 5.66   A view of Enceladus, showing a globe with densely cratered regions cut by areas of terrain with fewer, or no, craters.

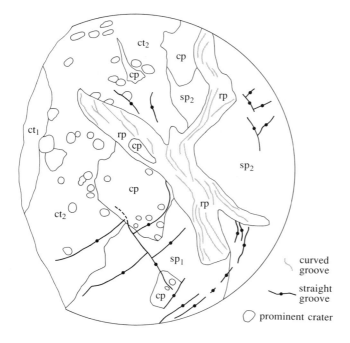

Figure 5.67   An interpretation of the image of Enceladus in Figure 5.66. The surface has been divided into six terrain units with different characteristics: ct_2, well-preserved fairly densely cratered; ct_1, similar to ct_2 but with highly flattened craters; cp, cratered plains (lower crater density than ct_1 or ct_2); sp_1, smooth plains with a few craters; sp_2, smooth plains with no visible craters; rp, ridged plains with no visible craters.

The geological history of Enceladus is therefore evidently long and complex. It is also poorly understood. Some areas of the lithosphere (ct_1 and ct_2) have evidently remained essentially undisturbed for a very long time, except for degradation of the crater morphology. The smooth plains (sp_1) were probably flooded cryovolcanically, removing evidence of the earlier craters, so that the only craters visible are those that were formed after the volcanic event. Volcanism occurred even more recently in the other terrain units, and there is good evidence of relatively young tectonic events also.

This history probably reflects several episodes of high heat flow, which appear to have affected different areas at different times. The best explanation for the episodicity is that the gravitational interactions between Saturn's satellites (which, after all, form a very complex system) have varied through time, so that Enceladus has experienced intervals when the forced eccentricity of its orbit was low (as at present) interspersed with periods of higher eccentricity when tidal dissipation would have been greater. The patchy distribution of the surface effects could either be a direct reflection of the scale of zones of convective upwelling in the asthenosphere due to this internal heating, or be a result of pre-existing zones of weakness in the lithosphere. Of course, none of this accounts for why such events occurred only on Enceladus and not on Mimas as well, and it is clear that we still have much to learn.

The division of Enceladus's surface into ancient, stable blocks cut by younger, more heavily deformed zones may remind you of the Earth's crust in the Archaean (Section 5.2, Figure 5.6). How close an analogue this is in terms of lithospheric and asthenospheric processes has yet to be established, but it is a good example of how the study of other bodies can make us think about how the Earth works, and what may have happened here in the past.

5.11.4   ARIEL

As our final example, we will look at Ariel, a satellite of Uranus. There are no orbital resonances amongst the Uranian satellites at present, but the evidence of past activity on Ariel suggests that there probably were once. Figure 5.68 (opposite) shows a general view of Ariel, and an enlargement of part of it is reproduced in Figure 5.69 (overleaf).

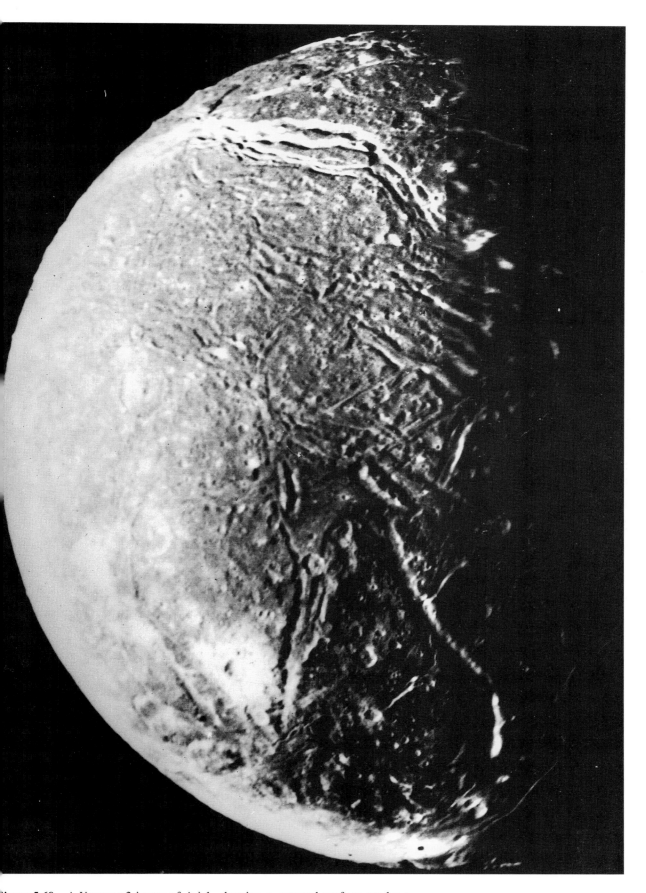

Figure 5.68 A *Voyager 2* image of Ariel, showing a cratered surface cut by a number of fault-bounded troughs.

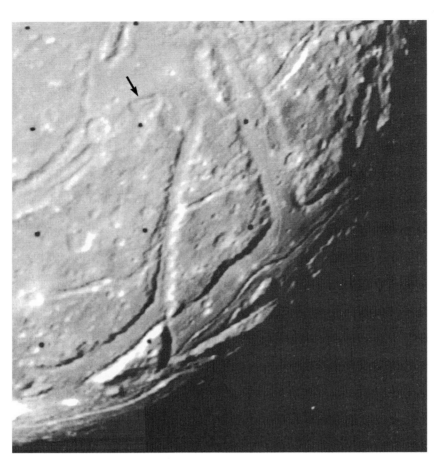

Figure 5.69 An enlargement from the lower part of Figure 5.68. The fault-bounded troughs are filled by a smooth, flow-like unit. Note also the site marked by an arrow, where this flow material floods half of a 30 km diameter impact crater.

Ariel's surface is cut by an extensive array of fault-bounded troughs. These seem to indicate fracturing of the lithosphere with limited motion between the resulting blocks, and may simply be due to to tidal stresses in the past. The troughs are filled by a smooth material that in places spills out beyond their ends to flood a broader area. The thickness of the flow (>1 km) and the steepness of its margin where it half-floods the crater indicated in Figure 5.69 show that the flow behaved very viscously, certainly far more viscously than water or any brine, and that it spread out far less readily than a basaltic flow would have done on the Earth. This is in marked contrast to the boundary between the ridged plains and cratered terrain on Enceladus (Figure 5.66), which has no visible topographic step.

The usual explanation is that the ice in Ariel (and the other satellites of Uranus) is not pure H_2O, and so has rather different properties. There is little direct proof of the nature of the contaminants, but it is likely that, with outwardly decreasing temperatures in the solar nebula, ammonia, methane, and ultimately nitrogen would become trapped in the phases that condensed. It is doubtful whether there is a significant amount of ammonia within Saturn's satellites, but there is good spectroscopic evidence of nitrogen on Neptune's largest satellite, Triton. The satellites of Uranus probably contain both ammonia and methane, in addition to salts of various kinds if there has been opportunity for water–rock reactions.

In order to understand cryovolcanism on Ariel more clearly, we will look simply at the effects of ammonia in ice. Within ice, ammonia molecules (NH_3) will not mix with water molecules (H_2O) in a solid solution series. Instead (for <50% NH_3 and at low pressure), there are only two possible solid phases, H_2O (pure water crystals) and $NH_3.H_2O$ (ammonia hydrate crystals), so the ice in Ariel must consist of interlocking crystals of these two phases analogous to an igneous or metamorphic silicate rock on Earth. Crystals of water and ammonia hydrate constitute a binary eutectic system analogous to the forsterite–diopside (Block 3, Section 3.4.2, Figure 3.23) and diopside–anorthite (Block 4, Section 4.2.4, Figure 4.8) systems, as shown in Figure 5.70.

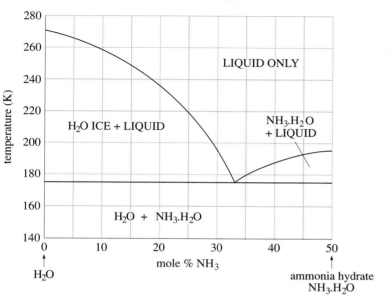

Figure 5.70 Phase diagram for the water (H_2O)–ammonia hydrate ($NH_3.H_2O$) binary eutectic system. Note that the horizontal axis is labelled in mole %, not mass %, which gives a direct representation of the relative numbers of molecules of H_2O and NH_3.

ITQ 41

Cosmochemical models suggest that ammonia is unlikely to make up more than about 20% of the total mass of the ice in any satellite. Assuming that the ice in Ariel is 20% ammonia and 80% water (i.e. 40% ammonia hydrate crystals and 60% water ice crystals), at what temperature would it begin to melt, and what would be the composition of the melt? (*Note:* the molecular masses of water and ammonia are 18 and 17 respectively, so 20% by mass of ammonia is very close to 20% (in fact 21.2%) by molecule)

Thus, a mixture of ammonia and water ices undergoes partial melting at a considerably lower temperature than either pure water or pure ammonia hydrate ($NH_3.H_2O$). This means that a much smaller amount of internal heat production is necessary for cryovolcanism to occur than in a body consisting of pure water ice. The ammonia–water eutectic melt is also less dense than that of ammonia-bearing ice, and so would rise buoyantly even without the aid of gas bubbles in the melt, or rock to weigh down the ice. Equally importantly, such a melt would have the sort of high viscosity necessary to explain the morphology of the cryovolcanic flows we can see on Ariel's surface, such as in Figure 5.69.

We conclude by looking briefly at the relative mobilities of different kinds of 'lava'. In very simple terms, the ease with which a flow can spread out must depend on its viscosity. A melt of low viscosity will be able to spread out thinly, but a high-viscosity melt of the same volume will spread less far and therefore produce a thicker flow. The exact form of a flow depends also on the rate at which it was erupted, but by looking at the thickness of a solidified flow (which we can measure at its edges) we can get an approximate idea of how mobile the melt was while it was in motion. However, to compare flows in this way on different planetary bodies, we must take into account the effect of different surface gravities, because on a high-gravity world a flow will spread further (and so be thinner) than on a low-gravity world. One approach to this problem is to calculate the **mobility index** (*I*) for the fluid in question, which provides a gravity-scaled measure of the fluidity of a flow. Mobility index is defined as follows:

$$I = \frac{\rho g}{\eta}$$ (Equation 5.6)

where ρ is the density of the melt ($kg\,m^{-3}$), g is the surface gravity of the planetary body ($m\,s^{-2}$) and η is the dynamic viscosity of the fluid ($kg\,m^{-1}\,s^{-1}$; see Block 1, Section 1.14.2). Put simply, this relationship means that a melt of a given viscosity will be less mobile (have a lower mobility index) on a world with a lower surface gravity. It is the mobility index of a melt that dictates the type of flow it will produce, so if we want to compare the behaviour of different kinds of melts on different planetary bodies then we have to calculate I.

Table 5.7   For completion in ITQ 42.

Melt	Dynamic viscosity, η ($kg\,m^{-1}\,s^{-1}$)	Density, ρ ($kg\,m^{-3}$)	Surface gravity, g ($m\,s^{-2}$)	Mobility index, I ($m^{-1}\,s^{-1}$)
Ammonia–water eutectic on Ariel	1×10^2	0.95×10^3		
Basalt on Earth	2.7×10^2	2.7×10^3	9.8	
Andesite on Earth	2×10^3	2.5×10^3	9.8	
Rhyolite on Earth	2.4×10^5	2.4×10^3	9.8	

ITQ 42

(a)   Ariel's mass is 13.5×10^{20} kg and its radius is 579 km (Table 5.2). Use Equation 1.34, $g = GM/d^2$, to calculate the surface gravity on Ariel. ($G = 6.67 \times 10^{-11}\,m^3\,s^{-2}\,kg^{-1}$.) Write this value in the appropriate cell in Table 5.7.

(b)   Now calculate the mobility index for an ammonia–water eutectic melt on Ariel, using the values of density and dynamic viscosity given in Table 5.7. Write your answer in the appropriate cell in Table 5.7.

(c)   Now calculate the mobility index for each of basalt, andesite and rhyolite on the Earth, writing your answers in Table 5.7.

(d)   On the basis of the mobility indices you have calculated, with which composition of lava flows on Earth is an ammonia–water eutectic flow on Ariel most closely comparable?

The mobility index exercise in this ITQ should have shown you that it is no surprise that the flows on Ariel are so much thicker than terrestrial basalt flows, despite the fact that an ammonia–water eutectic melt is actually less viscous than basalt. Detailed studies of flow morphology (which we are not going to go into) have to take into account other factors such as the effusion rate, the minimum shear stress necessary before a real lava (silicate or cryovolcanic) will begin to flow, the ability of a chilled crust to confine a flow, and the presence of crystals within the melt. The latter two act so as to increase the effective viscosity of a flow, and therefore to decrease its ability to spread out.

SUMMARY OF SECTION 5.11

Europa has a thin ice lithosphere and a young surface, as a result of heating by tidal dissipation. It is the only reasonable candidate for a body other than the Earth where plate recycling has played a major role in outwards heat transfer during recent geological time.

Enceladus shows evidence of several episodes of volcanic and tectonic activity, possibly related to a variable history of internal heating by tidal dissipation. In contrast, its similarly sized neighbour, Mimas, has merely an ancient cratered surface with no signs of comparable activity.

The presumed incorporation of ammonia into the ice in satellites beyond Saturn, such as Ariel, enables partial melting yielding an ammonia–water eutectic melt (approximately 33% ammonia) at 100 K below the melting point of pure ice. Under Ariel's low surface gravity, the behaviour of this melt mimics that of a terrestrial magma intermediate between andesite and rhyolite in composition. This relationship can be expressed by the mobility index, $I = \rho g / \eta$.

OBJECTIVES FOR SECTION 5.11

When you have completed this Section, you should be able to:

5.1   Recognize and use definitions and applications of each of the terms printed in the text in bold.

5.25   Understand how differing compositions and contrasting thermal events can give rise to a diversity of histories among the larger (>200 km radius) icy satellites.

5.26   Have revised your ability to use and understand binary eutectic phase diagrams.

5.27   Be able to compare the mobility of melts on different planetary bodies, and understand why it is necessary to take into account both their viscosities and the surface gravity.

5.28   Be able to compare icy satellites with terrestrial planets in terms of what we know about them and how we know it.

Apart from Objective 5.1, to which they all relate, the five ITQs in this Section test the Objectives as follows: ITQs 38 and 41, Objective 5.26; ITQs 39 and 40, Objective 5.25; ITQ 42, Objective 5.27.

SAQS FOR SECTION 5.11

SAQ 30 (*Objective 5.25*)

Why is ammonia unlikely to play a role in cryovolcanism on Europa?

SAQ 31 (*Objective 5.27*)

Use the data in Table 5.2 (p. 41) to decide whether an ammonia–water eutectic melt would be more mobile or less mobile on Miranda than on Ariel.

SAQ 32 (*Objective 5.28*)

(a)   Our ideas about the internal structures of the planetary bodies in the Solar System are based on a wide range of data. The data are much more complete in some cases than others. Complete Table 5.8 by ticking the appropriate boxes for each body or class of bodies for which data exist. Put brackets around the ticks to represent cases where the data are notably poorer than for certain other bodies.

(b)   What type of data (as opposed to cosmochemical modelling) on the chemical composition of the icy satellites and Io is not included in this Table?

Table 5.8   For completion in SAQ 32.

	Chemical sampling	Seismic measurements	Moment of inertia measurements	Density determinations
Earth				
Moon				
Venus				
Mars				
Io				
Icy satellites				

5.12   WORLDS OF DIFFERENCE

In Section 5.6, we posed the question of whether a planet like the Earth must inevitably develop in the same way. By now, you should have reached the conclusion that the answer to this question is almost certainly no. Many bodies, whether terrestrial planets or the larger of the icy satellites, probably became differentiated into cores, mantles and crusts as a result of accretional heating, topped up by radiogenic heating if they contain substantial rocky fractions. Subsequent histories, though, are strongly dependent on a complex interplay of gross compositional effects, the role and location of volatiles (notably water in the case of Venus and ammonia in the case of Ariel), heating mechanisms, surface temperature, and other factors. In concert, these dictate the rate at which heat must be transferred outwards through the lithosphere, whether this occurs by plate recycling, volcanism or simple conduction, and how the balance between these processes changes with time.

If the Moon was formed as a result of a giant impact (Block 1, Section 1.3.1), then the Earth's balance of volatiles must have been profoundly altered, and its gross composition would have been changed to a lesser extent; consequently, it is unlikely that the Earth would resemble its present form had this event not occurred. Even if the giant impact hypothesis of lunar origin is wrong, then the Earth's evolution is still likely to have been subject to random factors. Here are two examples; doubtless you can think of others:

1   The amount of water originally in the outer part of the Earth would depend on how many of the last few impacts in the late heavy bombardment were by ice-rich comets or rock-rich asteroidal fragments, and this could have tipped the balance between the ultimate establishment of Earth-style or Venus-style tectonics. You saw in Section 5.8.4 that the ability of mantle melts in Venus to reach the surface may be dictated by the amount of water present.

2   The Moon is responsible for a small amount of tidal heating within the Earth. This would have been greater in the past, when the radius of the Moon's orbit was less (Block 1, Section 1.3.3). The amount of heating and its variation over time would depend on the mechanical properties of the interiors of both the Earth and the Moon, and on the Moon's mass. If the Moon's mass and properties were even slightly different (irrespective of how the Moon actually formed), it is conceivable that this could have set the Earth off on a different evolutionary path so that its pattern of tectonism and volcanism might now resemble that of Venus, Io, Enceladus or even Mars.

Speculation like this could go on indefinitely, and inconclusively, bearing in mind that (despite the impression you may have formed earlier in this Course) there is a great deal we still do not understand about the Earth itself, let alone its nearest neighbours. We hope by now that you have come to appreciate what a fascinating problem it is to attempt to unravel how the Earth (or any other planetary body) works.

ITQ ANSWERS AND COMMENTS

ITQ 1

You should recall from Block 2, Section 2.5.4, that the assembly of Gondwanaland is based on (i) palaeomagnetic data, notably polar wander curves, and (ii) sea-floor magnetic stripes, which can be used to back-track the opening of oceans, and from Block 2, Section 2.2, that the reconstruction of continents in general can be interpreted with the aid of (iii) geometrical fits and by matching distinctive geological formations across the joins (e.g. Figures 2.4 and 2.5). Other evidence (that you have not met in this Course) includes the fossil record, which demonstrates widespread common flora and fauna across Pangaea, that subsequently became regionalized into first Gondwanan and Laurasian types and then into groups characteristic of each continental fragment.

ITQ 2

North America and Europe were formerly separated (by a now-vanished ocean) which was closing during the period 550 Ma to about 390 Ma. The two continents remained joined until the present North Atlantic Ocean began to open about 150 Ma ago.

ITQ 3

As it ages, the oceanic lithosphere will cool and become progressively less buoyant (Block 1, Section 1.12.4). Thus, as time goes by, there will be a progressively greater tendency for subduction to be initiated at the hitherto passive margins of the Atlantic-type oceans. Once subduction has begun, the slab-pull force will tend to close the ocean.

ITQ 4

$$C_1 = \frac{C_0}{D + (1-D)F}$$

For Sr,

$$C_1 = \frac{260}{0.2 + (1 - 0.2)\,0.1}$$

$$= 929 \text{ ppm}$$

For Y,

$$C_1 = \frac{40}{10 + (1 - 10)\,0.1}$$

$$= 4.4 \text{ ppm}$$

Hence, Sr/Y = 211, Y = 4.4.

This point is plotted on the dashed line on Figure 5.7 which indicates the Sr/Y and Y values of melts for varying degrees of partial melting of eclogite. It appears that the great majority of Archaean data can be modelled from varying degrees of partial melting of an eclogite source. This suggests that during the Archaean eclogite rather than peridotite provided the source of most magmas involved in crustal growth. Eclogite is a metamorphic rock of basaltic composition that occurs in subducted oceanic lithosphere below destructive plate margins.

ITQ 5

We know from Block 4 (Figure 4.4) that only if the geothermal gradient is unusually high will the subducted slab be heated above its solidus. We have therefore arrived at the conclusion that during the Archaean, melts were predominantly derived from partial melting of eclogite and that geothermal gradients were higher than they are today.

This is an important conclusion because it means that the processes described in Block 4, Sections 4.2 and 4.3, for deriving magmas of intermediate composition from partial melting of the mantle (to produce basalt) and fractional crystallization of basalts (to produce andesite) were not operative throughout Earth history. Indeed, it seems that about 50% of continental crust was generated in the Archaean (Figure 5.6) by partial melting of the subducted slab (eclogite) to form magmas of intermediate composition.

It would, however, be a mistake to conclude that there was a dramatic shift from one mechanism to another at the end of the Archaean. Indeed, there is a small proportion of magmas formed at present-day island arcs with compositions that lie not in the high Y, low Sr/Y field of island arcs in Figure 5.7, but in the low Y, high Sr/Y field of Archaean igneous rocks. Island arcs that generate magmas of this composition share one feature in common; the age of subducted lithosphere is unusually young (< 25 Ma) in each case. One example is the Aleutian arc which stretches southwestwards from Alaska as you can find on the Smithsonian Map. The young age of the subducted lithosphere in this case (~ 10 Ma) results from subduction of a RRR triple junction, as you may recall from Block 3 (Section 2.3.5). The occurrence of such geochemical characteristics in recent island arcs suggests that even today, melting of the subducted slab *can* occur but the lithosphere must be young for that to be possible. Since oceanic lithosphere cools gradually after formation at an ocean ridge (Block 1, Section 1.12), this means that slab melting at destructive margins today will be possible *provided* the subducted slab is young, and therefore hot. This is not the case for the majority of modern destructive margins (including the Andes) and so it appears that since the Archaean there has been a gradual change from melting of the subducted slab to melting in the mantle wedge which predominates today.

ITQ 6

From the definition of the Urey ratio (Equation 5.2), we have $Ur = 2.2 \times 10^{13}/4 \times 10^{13} = 0.55$.

ITQ 7

From Equation 5.5:

[handwritten annotations: *Earth mass*, *Specific Heat Capacity*]

2.4×10^{13} J s^{-1} = 4.06×10^{24} kg $\times$ 1 000 J kg^{-1} K^{-1} $\times$ rate of change in temperature (T')

$T' = (2.4 \times 10^{13} \text{ J s}^{-1})/(4.06 \times 10^{24} \times 10^3 \text{ J K}^{-1})$

$T' = (2.4 \times 10^{13} \text{ J s}^{-1})/(4.06 \times 10^{27} \text{ J K}^{-1})$

$T' = 5.91 \times 10^{-15}$ K s^{-1}

$T' = 5.91 \times 10^{-15} \times 3.15 \times 10^7$ K a^{-1}

$T' = 5.91 \times 10^{-15} \times 3.15 \times 10^7 \times 10^9$ K Ga^{-1}

$T' = 186$ K Ga^{-1}

ITQ 8

Given the MgO contents in Table 5.1, the 1 atmosphere liquidus temperatures read off from Figure 5.11 are 1: 1 660 °C, 2: 1 445 °C, 3: 1 425 °C, and 4: 1 250 °C. These temperatures are reported to the nearest 5 °C, reflecting uncertainties in reading the graph.

ITQ 9

No. The highlands region shows many times more craters than the mare, despite being only about 25% older. This shows that the rate of impact cratering experienced by the Moon must have declined markedly from a high rate to a much lower rate some time between 4 420 Ma and 3 500 Ma ago.

ITQ 10

The conversion of kinetic energy to heat upon impact will have caused accretional heating (Block 1, Section 1.3.3). The effect this will have had depends on the depth to which the incoming heat was mixed. However, it seems reasonable that, by maintaining the near-surface region at a higher temperature than it would otherwise have experienced, the late heavy bombardment would, on average, make the geothermal gradient (as defined in Block 1, Section 1.12.2) less steep. Of course, immediately following an impact, the temperature might be so elevated near the surface that temperature could actually decrease with depth, leading to a (short-lived) reversed geothermal gradient. Note that this discussion is equally relevant to the (undocumented) period prior to the late heavy bombardment, when the rate of impact cratering was probably just as great, if not greater.

ITQ 11

The obvious ones are the planets whose size and density most resemble the Earth's; these are the inner planets: Mercury, Venus and Mars. They are dense enough to be essentially rocky but with an even denser core rich in iron and nickel. The Moon is also a sensible candidate, not just because it is the one we know most about (having been investigated by several manned and robotic missions, many of which have landed on its surface), but also because its density and composition make it comparable with the inner planets.

ITQ 12

Block 1, Table 1.1, shows that the Moon is both the smallest and the least dense of the terrestrial planets. According to Table 5.2, three of the four major satellites of Jupiter, and Titan, the largest satellite of Saturn, are bigger than the Moon. The radius of Europa, the remaining large satellite of Jupiter, is only 10% less than that of the Moon, and that of Triton, the major satellite of Neptune, is 23% less. All the other planetary satellites are very much smaller. In terms of density, only Io lies within the range spanned by the traditional terrestrial planets, having a density intermediate between that of the Moon and Mars. The density of Europa is only about 10% less than that of the Moon, but all the others are considerably less dense. You may have spotted that inspection of the data in these Tables confirms that Pluto, whose radius is about 1 142 km and whose mass is about 124×10^{20} kg, seems to bear more resemblance to Triton (the major satellite of Neptune) than to any of the other planets.

ITQ 13

The best way to work this out is by simple proportions: if the fraction (by volume) of Enceladus that is composed of rock is expressed by x, then the fraction by volume composed of ice must be $(1-x)$. You can then construct the expression $\rho = x\rho_{rock} + (1-x)\,\rho_{ice}$, where ρ is the average density, ρ_{rock} is the density of rock, and ρ_{ice} is the density of ice.

Inserting the values given in the question, this becomes:

$$1.2 \times 10^3 \text{ kg m}^{-3} = x\,(3.0 \times 10^3 \text{ kg m}^{-3}) + (1-x)\,(0.95 \times 10^3 \text{ kg m}^{-3})$$

which can be rearranged to give:

$$(1.2 - 0.95) \times 10^3 \text{ kg m}^{-3} = x\,(3.0 - 0.95) \times 10^3 \text{ kg m}^{-3}$$

which reduces to $0.25 = 2.05x$, so $x = 0.25/2.05 = 0.12$.

Thus, the fraction by volume of Enceladus made of rock is 0.12 (i.e. 12%), if the remainder of the body (88%) is composed of ice.

(*Comment:* If you are more interested in the fraction by *mass*, M_{rock}/M, where M_{rock} is the mass of rock and M is the total mass, then the conversion is as follows: $x = V_{rock}/V$, where V_{rock} is the volume of rock and V is the total volume, but $V_{rock} = M_{rock}/\rho_{rock}$, and $V = M/\rho$.

So, $x = (M_{rock}/\rho_{rock})/(M/\rho) = M_{rock}\rho/M\rho_{rock}$, and hence $M_{rock}/M = x\rho_{rock}/\rho$. Inserting values for this case, $M_{rock}/M = 0.12 \times 3.0/1.2 = 0.3$. It would also be possible to calculate M_{rock}/M directly, without calculating V_{rock}/V first.)

ITQ 14

Since the Moon's rotation on its axis keeps pace with its orbital revolution, it must rotate exactly once during each orbit; indeed, you were told as much in Block 1, Section 1.2.1. If you have difficulty visualizing this, look at Figure 5.71 which explains it. You might like to try looking at the Moon on successive nights to convince yourself that this is correct: although the portion illuminated by the Sun (and therefore visible) changes from night to night, the surface markings remain fixed.

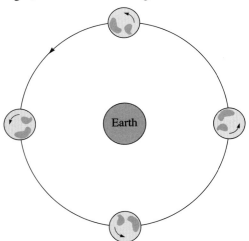

Figure 5.71 Synchronous rotation of the Moon. The Figure shows a view of the Earth–Moon system seen from above (north of) the plane of the Moon's orbit (not to scale), with the Moon shown at four successive positions. Schematic surface features are shown on the Moon for reference. By the time the Moon has completed a single orbit about the Earth it has rotated exactly once on its axis. In consequence, the Moon keeps the same face pointing towards the Earth at all times.

ITQ 15

Because the tidal bulges raised by the Earth on the Moon are stationary, no energy is expended in flexing the body of the Moon, so there is no resulting generation of heat. (*Comments*: Actually, the Moon's orbit is not quite circular, so the tidal bulge changes in height and position slightly during the course of an orbit, but the rate of heat generation is negligible. The Sun raises tides in the solid Moon, too, and, in contrast to the tides raised by the Earth, these must propagate round the Moon so as to keep pace with its rotation relative to the Sun; however, the rate of heating attributable to this is very small.)

ITQ 16

(a)   The Earth's global heat production is given by its surface area times its heat flow per unit area ($80\,\text{mW}\,\text{m}^{-2} = 0.08\,\text{W}\,\text{m}^{-2}$), i.e.

$$5.1 \times 10^{14}\,\text{m}^2 \times 0.08\,\text{W}\,\text{m}^{-2} = 4.1 \times 10^{13}\,\text{W}.$$

Using the same equation, the Moon's global heat production is:

$$3.8 \times 10^{13}\,\text{m}^2 \times 0.015\,\text{W}\,\text{m}^{-2} = 5.7 \times 10^{11}\,\text{W}.$$

(b)   The Earth's mass is approximately $6.0 \times 10^{24}\,\text{kg}$, so its rate of heat generation per kg is global rate of heat production/mass:

$$(4.1 \times 10^{13}\,\text{W})/(6.0 \times 10^{24}\,\text{kg}) = 6.8 \times 10^{-12}\,\text{W}\,\text{kg}^{-1}.$$

The Moon's mass is approximately $7.3 \times 10^{22}\,\text{kg}$, so its rate of heat generation per kg is:

$$(5.7 \times 10^{11}\,\text{W})/(7.3 \times 10^{22}\,\text{kg}) = 7.8 \times 10^{-12}\,\text{W}\,\text{kg}^{-1}.$$

ITQ 17

Table 5.3 shows clearly that all four siderophile elements listed have lower concentrations in the Moon than in the Earth, by about one to three orders of magnitude, thus the Moon is depleted in siderophile elements relative to the Earth's mantle and crust.

ITQ 18

The first thing we need to do is to recast the concentrations in Table 5.3 in terms of heat-producing isotopes. Dividing U according to the percentages given in the question, of the total of 18 ppb U in the Earth's mantle plus crust, 0.1 ppb must be ^{235}U and 17.9 ppb must be ^{238}U. The concentration of K is given as 180 ppm, and as only 0.01% (one part in 10^4) of this is ^{40}K the concentration of ^{40}K must be 18 ppb.

Following the method of Block 1, ITQ 48(a), radiogenic heat production per kg = mass of isotope per kg of material × heat generation by 1 kg of that isotope.

For ^{235}U, this is:

$$0.1 \times 10^{-9} \times 56 \times 10^{-2}\,\text{mW}\,\text{kg}^{-1} = 0.056 \times 10^{-9}\,\text{mW}\,\text{kg}^{-1}$$

For ^{238}U, this is:

$$17.9 \times 10^{-9} \times 9.6 \times 10^{-2}\,\text{mW}\,\text{kg}^{-1} = 1.72 \times 10^{-9}\,\text{mW}\,\text{kg}^{-1}$$

For ^{232}Th, this is:

$$80 \times 10^{-9} \times 2.6 \times 10^{-2}\,\text{mW}\,\text{kg}^{-1} = 2.08 \times 10^{-9}\,\text{mW}\,\text{kg}^{-1}$$

For ^{40}K, this is:

$$18 \times 10^{-9} \times 2.8 \times 10^{-2}\,\text{mW}\,\text{kg}^{-1} = 0.504 \times 10^{-9}\,\text{mW}\,\text{kg}^{-1}$$

Summing these values, the predicted rate of radiogenic heat production from the Earth's mantle plus crust, according to the estimates of composition in Table 5.3, comes to $c.\ 4.4 \times 10^{-9}\,\text{mW}\,\text{kg}^{-1}$. This is slightly lower than the $5 \times 10^{-9}\,\text{mW}\,\text{kg}^{-1}$ calculated for chondritic material in Block 1, ITQ 48(a), but indistinguishable from the $4 \times 10^{-9}\,\text{mW}\,\text{kg}^{-1}$ estimated for the Earth in part (b) of that question (this is no surprise, as the heat generation values for the Earth's crust and mantle used in the question were based on geochemical data like those in Table 5.3).

For the Moon, the concentrations of the heat-producing isotopes can be derived from Table 5.3 in the same manner as for the Earth. The 33 ppb U must be 0.2 ppb ^{235}U and 32.8 ppb ^{238}U, whereas the concentration of ^{40}K must be 8.3 ppb.

We can then calculate the radiogenic heat production per kg of Moon, in the same way as for the Earth:

For ^{235}U, this is:

$$0.2 \times 10^{-9} \times 56 \times 10^{-2} \text{ mW kg}^{-1} = 0.112 \times 10^{-9} \text{ mW kg}^{-1}$$

For ^{238}U, this is:

$$32.8 \times 10^{-9} \times 9.6 \times 10^{-2} \text{ mW kg}^{-1} = 3.15 \times 10^{-9} \text{ mW kg}^{-1}$$

For ^{232}Th, this is:

$$112 \times 10^{-9} \times 2.6 \times 10^{-2} \text{ mW kg}^{-1} = 2.91 \times 10^{-9} \text{ mW kg}^{-1}$$

For ^{40}K, this is:

$$8.3 \times 10^{-9} \times 2.8 \times 10^{-2} \text{ mW kg}^{-1} = 0.232 \times 10^{-9} \text{ mW kg}^{-1}$$

Summing these values, the predicted rate of radiogenic heat production from the Moon's mantle plus crust, according to the estimates of composition in Table 5.3, comes to approximately 6.4×10^{-9} mW kg^{-1}.

ITQ 19

The simplest to interpret is above A, which is clearly the megaregolith, where seismic velocity is low. The major velocity increase at C is universally recognized to be the base of the crust, so below C is the mantle; note that a P-wave velocity of around 9 km s^{-1}, like that below C, is similar to that in the Earth's upper mantle (Block 1, Figure 1.52). If, like the lunar scientists who originally worked on these data, you thought that the region from A to B represents the mare basalts and that the velocity increase at B represents the underlying highland crust, you cannot be blamed. However, a variety of evidence suggests that mare basalts are generally much thinner than the 20 km that this interpretation would require, and moreover there appears to be a similar velocity discontinuity below the *Apollo 16* site, which is actually on the highlands. The best interpretation is that the change between mare basalt and highland crust occurs at most a few km below A, and that it is masked by a gradual increase in velocity caused by the closure of impact-generated cracks with increasing pressure. The sharp jump in velocity at B probably represents a compositional change within the highland crust, perhaps to a more Mg-rich variety of anorthosite, and we can draw analogies between this and seismic continuities in the Earth's continental crust such as those shown in Block 1, Figure 1.58.

ITQ 20

If S-waves are not transmitted at all, this would mean that the medium had become a liquid, perhaps like the Earth's outer core. However, by now you should rule this out straight away, on the grounds of the Moon's low density and its lack of a significant magnetic field. If, however, the lack of detectable S-wave propagation is simply due to attenuation, then it suggests that this depth represents the start of the lunar asthenosphere, which could be partially molten. The coincident drop in P-wave velocity suggests that 1 000 km can be taken as the base of the Moon's seismic lithosphere, according to the definition given in Block 1, Section 1.13. Incidentally, no P-wave reflections have been detected from this boundary, which has been used to argue that it is a diffuse transition, rather than a sharp interface.

ITQ 21

They coincide with the sites of several of the more circular maria, namely Mare Imbrium, Mare Serenitatis, Mare Crisium, Mare Humorum and Mare Nectaris.

ITQ 22

The mascons demonstrate that isostatic equilibrium has not been attained (e.g. Block 1, Figure 1.74). In the Earth, the level of isostatic compensation is usually taken to be the top of the asthenosphere (Block 1, Section 1.8.5). On the Moon, the crust and the underlying mantle lithosphere are evidently strong and rigid, and have been so at least since the time of the mare-filling events. This is entirely compatible with the present-day 1 000 km lithospheric thickness shown in Figure 5.30 (which comprises the crust and all the mantle above the asthenosphere).

ITQ 23

The Venus data bear a close resemblance to oceanic basalts, and are quite clearly different to Earth's continental crust, having too little SiO_2 and too much MgO, FeO and CaO. Note however that the *Venera 13* and *Vega 2* analyses fall on the 45% SiO_2 division between ultrabasic and basic composition (Block 3, Appendix 2) and have MgO contents approaching those of komatiites (Section 5.4.3).

ITQ 24

(a)   Your completed plot should look like that in Figure 5.72.

(b)   There is only one peak in the Venus data (i.e. the distribution is unimodal), so Venus does not appear to show an Earth-like distinction of two types of crust ('continental' and 'oceanic'). Note, however, that this comparison may be somewhat flawed by virtue of the very different surface processes that operate on the two planets. The Earth's mountainous regions suffer erosion primarily by flowing water and ice, which transport the resulting sediment downhill and deposit it near sea-level, thereby exaggerating the peak near sea-level on Figure 5.34. This cannot happen on Venus, which lacks flowing water and vegetation to bind the loose sediment. Nevertheless, it is difficult to see how one could go from a unimodal, Venus-like, hypsographic curve to a bimodal, Earth-like, curve simply by invoking erosion and deposition, and most planetary scientists would agree that the data in Table 5.5 do indicate a fundamental difference between the crusts of the Earth and Venus.

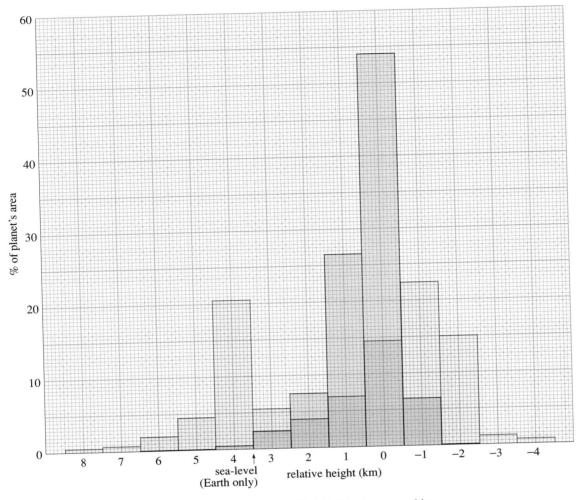

Figure 5.72 Completed version of Figure 5.34 (for ITQ 24). The hypsographic plot for Venus is shown in black.

ITQ 25

The resemblance to oceanic plate tectonics is slight. No spreading axes or transform faults have been identified (though there are a few trenches that may mark subduction zones, as noted at the end of Section 5.8.2). The pattern of deformation (broad deformed belts separating more stable blocks) is more similar to that within the Earth's continents (as summarized in Block 2, Section 2.5), with the exception that there appear to be no major strike–slip features comparable to the San Andreas Fault.

ITQ 26

The discussion at the end of Section 5.5 ought to make it clear to you that the scarcity of craters on the surface of Venus is good evidence that, on the whole, it is considerably younger than the lunar maria. In fact, the best age estimate for most of the surface based on crater counts is that it spans a range from approximately 0 to 800 Ma, with an average age of about 500 Ma. It is worth noting that on Venus impact craters smaller than 35 km are rare, whereas the abundance of craters larger than this follows the same shaped size-frequency distribution as lunar and martian craters. This is consistent with calculations suggesting that the atmosphere should have no effect in impeding the flux of large impactors reaching the surface, but that it is sufficiently dense that smaller impactors (that would have produced craters less than 35 km in diameter) burn up before reaching the surface.

ITQ 27

Depth is related to geothermal gradient by: $\text{depth} = \dfrac{\Delta T}{\text{geothermal gradient}}$

(a) Substituting values for the Earth, using ΔT as the difference between the melting temperature of anhydrous granulite and the surface temperature:

$$\text{depth} = \frac{1\,100 - 20\,°C}{25\,°C\,\text{km}^{-1}} = \frac{1\,080\,\text{km}}{25} = 43.2\,\text{km}$$

(which is within the range of actual values of continental crustal thickness encountered in Block 1, Section 1.7.2).

(b) Substituting values for Venus:

$$\text{depth} = \frac{1\,100 - 457}{25\,°C\,\text{km}^{-1}} = \frac{643\,\text{km}}{25} = 25.7\,\text{km}$$

(*Comment*: There are several limitations inherent in this simple model. One is that heat-producing elements in Venus are likely to be concentrated in the crust (as on Earth), so the geothermal gradient must decrease deeper in the crust. In the case of the Earth's continents, this is exacerbated because about 70% of heat flow is generated within the upper 10–20 km (Block 1, Sections 1.12.2 and 1.12.5). However, it seems clear that Venus's high surface temperature means its crust must be thinner than the Earth's.)

ITQ 28

Following the logic of Block 3, Section 3.5.1, and treating the lithosphere as a conductive boundary layer, the part of the geothermal gradient where temperature increases most rapidly with depth is the conductive lithosphere. Below this, where temperature increases less rapidly with depth (at about 100 km beneath old oceanic crust in the Earth according to Block 3, Figure 3.31) is the convecting regime (where the temperature gradient is adiabatic), and this may be regarded as the asthenosphere. The change in gradient in Figure 5.41 is at a pressure of about $1.5\,\text{GN}\,\text{m}^{-2}$. Given that $3.0\,\text{GN}\,\text{m}^{-2}$ corresponds to 105 km on Venus, $1.5\,\text{GN}\,\text{m}^{-2}$ implies a depth of approximately 50 km, and so the lithosphere of Venus would appear to be only about 50 km thick, or only about half the thickness of the thermal lithosphere on Earth (Block 1, Section 1.13).

ITQ 29

The hypsographic plot for Mars is neither strongly bimodal like that for for the Earth nor, strictly, unimodal like that for Venus (see Figure 5.72). It shows a wider range of heights than either of the other planets, with several subsidiary peaks.

ITQ 30

By cancelling out g, the expression, $t\rho_r g = (s + t)\rho_m g$ becomes:

$t\rho_r = (s + t)\rho_m$, so $s\rho_m = t\rho_r - t\rho_m = t(\rho_r - \rho_m)$, so

$$t = \frac{s\rho_m}{(\rho_r - \rho_m)}.$$

Substituting the values given, this becomes:

$$t = \frac{24\,\text{km} \times 2.8 \times 10^3\,\text{kg}\,\text{m}^{-3}}{3.2 \times 10^3\,\text{kg}\,\text{m}^{-3} - 2.8 \times 10^3\,\text{kg}\,\text{m}^{-3}} = \frac{2.8 \times 24\,\text{km}}{3.2 - 2.8} = \frac{2.8 \times 24\,\text{km}}{0.4}$$

$= 168\,\text{km}$, or approximately 170 km.

ITQ 31

(a)   If we accept that Io and the Moon have approximately equal masses of silicates, and that the role of heat-producing elements in the core is minor, then there should be similar rates of radioactive heat production in both bodies. (The uncertainties over the comparison between radiogenic heat production in the Earth and the Moon that you encountered in ITQs 16 and 18 should make you wary of too close a reliance on the assumption that the radioactive elements are in equal abundance in both bodies, but there is no reason to expect Io to have a *dramatically* different concentration of radioactive elements to the Moon.)

(b)   Io's slightly larger size than the Moon means that its surface area is greater, allowing it to lose heat by radiation more efficiently than the Moon. Thus, if Io produces heat at the same rate as the Moon, it ought, if anything, to be a slightly colder, less active body than the Moon. As you learned in Section 5.7, the Moon shows no clear signs of volcanic activity over the past 3 billion years, and so it is extremely unlikely that radioactive heat production within Io could give rise to any present-day volcanic activity. (*Comment*: The amount of heat generated during accretion may well have been different for each body, but this would be unlikely to lead to much of a difference in the present-day heat flows.)

ITQ 32

The thing that should strike you in particular is that Io appears devoid of impact craters (in fact, none has been identified on any of the *Voyager* images). This means that Io's surface must be relatively young, certainly compared to the Moon, Venus, and Mars (and most of the other outer planet satellites, as you will see in Section 5.11). There are some depressions on Io's surface, but these appear volcanic in nature, being less circular than impact craters and lacking their characteristic central peaks and surrounding ejecta blankets. In addition, there are several features that could be lava flows covering parts of the surface (indicated on Figure 5.73), so in addition to being resurfaced by fall-out from eruption plumes, at least parts of Io are covered by lavas of some kind.

Figure 5.73   Annotated version of Figure 5.57, showing lava flows and volcanic calderas.

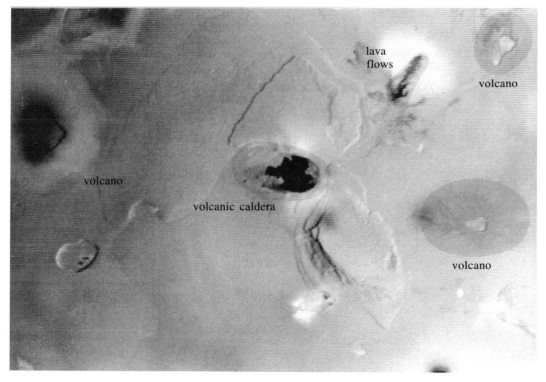

TQ 33

The Moon's heat flow is about 0.015 W m^{-2} (p. 46), and at 1–3 W m^{-2} Io's heat flow is some 65–200 times greater, despite being closely similar in size and mass (for comparison, this is about 16–50 times greater than the Earth's average heat flow).

TQ 34

Io's orbital period is almost exactly half that of Europa, and a quarter that of Ganymede.

TQ 35

Io's rotation cannot be exactly synchronous. Although Io rotates exactly once in every orbit, the changing speed of its orbital motion means that the rotation gains slightly on the orbital motion when Io is furthest from Jupiter, and lags behind slightly when Io is closest to Jupiter. This means that, from the point of view of an observer on Jupiter, Io would appear to wobble slightly. We hope that Figure 5.74 will make this clear to you.

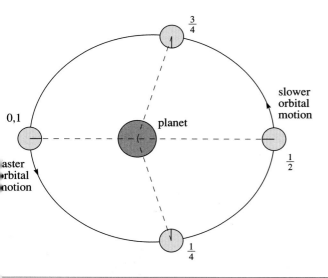

Figure 5.74 The effect of an elliptical orbit on a synchronously rotating satellite. The satellite is shown at a starting position when it is closest to its planet and three subsequent positions after a quarter, half, and three-quarters of its orbital period have elapsed. The dashed lines indicate the line joining the centres of the planet and satellite in each position, and the solid line drawn on the satellite is a reference line to show how far it has rotated. A tidal bulge will try to maintain its position on the satellite's surface centred on the dashed line, which means it must move relative to the reference line. (The shape of the orbit has been exaggerated for clarity.)

TQ 36

The locations of the tidal bulges will have to oscillate to keep pace with the libration, so as to keep the mid-points of the bulges exactly on a line between the centres of Io and Jupiter.

TQ 37

Io's lithosphere is shown about 40–50 km thick in Figure 5.59, whereas you know from Block 1, Sections 1.7.5 and 1.13, that the Earth's lithosphere is about 30–60 km thick under much of the oceans and somewhat over 100 km thick under the continents. The thickness is greater if we define the lithosphere thermally (75–125 km in the oceans). If the asthenosphere in Figure 5.59 is defined to be a convecting layer, then it is effectively Io's *thermal* lithosphere that is indicated on this Figure. Of the other terrestrial planets you have studied in this Block, Venus's thermal lithosphere is probably about 50 km thick (ITQ 28) though its elastically and seismically defined lithospheres are probably over twice this thickness (Section 5.8.4), the Moon's is 1000 km or more in thickness (Section 5.7.3), and Mars's is probably about 170 km thick (Section 5.9.2).

ITQ 38

The first melt is represented by the eutectic point, i.e. the minimum point between the two liquidus curves. You should realize this by analogy with other binary eutectic systems you have met, such as forsterite–diopside (Block 3, Section 3.4.2) and diopside–anorthite (Block 4, Section 4.2.4). In this example, the eutectic point is at a temperature of $-3.5\,°C$ and has a composition of 17 mass % $MgSO_4$ (the remaining 83 mass % being water).

ITQ 39

(a) (i) The surface must be very young; much younger than the lunar maria for example (in fact, its average age is unlikely to be older than 100 Ma, and could be much younger). (ii) The low relief implies that Europa's lithosphere is probably too thin and weak to support the weight of major topographic features.

(b) These both show that Europa has recently been (and probably still is) considerably more geologically active than, say, the Moon or Mars. The young surface is almost certainly the result of cryovolcanism.

ITQ 40

Note: The following is rather a fuller answer than we would have expected from you, but we hope that you got the main points, and, having studied our answer would be able to make an equally full interpretation of the surface of any other planetary body.

The heavy cratering of ct_1 and ct_2 indicates that these regions have surfaces dating back a long time (possibly to about the end of the late heavy bombardment, though as remarked in Section 5.5 the lunar cratering time-scale is probably invalid in the outer Solar System). The cratered plains (cp) have a lower density of craters, and so are younger, sp_1 has even fewer craters and so must be younger still, and sp_2 and rp have no visible craters and so are the youngest regions. Thus, the time-span of the activity appears to be very long. The flattening of the craters in ct_1 suggests either isostatic adjustment of the craters (in a thin but rigid lithosphere) or gradual creep of the surface ice under its own weight (suggesting the ice became relatively warm and mushy at some time after crater formation). Some of the larger craters in ct_2 also have strange morphologies, particularly the prominent pair near the ct_1/ct_2 boundary whose central peaks are unusually large. You should also have noticed some craters with half missing (at and near the boundary between rp and ct_2); the likely explanation here is faulting and subsidence followed by flooding (by water or mushy ice) that buried the older surfaces. The straight grooves are probably signs of tectonic activity; some grooves cutting sp_1, cp and ct_2 meet at right angles in the lower part of the Figure, and the apparent offset of one groove by another suggests possible sideways movements. It is tempting to regard the curved grooves in rp as some kind of compressional feature (folds perhaps), but they could equally well be some kind of extrusional (cryovolcanic) landform. The truncation of older craters by the ridged plains, noted earlier, favours the extrusional interpretation.

ITQ 41

The first melt would form at the eutectic point, which in this system is at 176 K and with a composition of about 33% ammonia and 67% water. Pure H_2O ice would remain as a solid phase. If you could not answer this, you should refer back to the discussion of the forsterite–diopside system in Block 3, Section 3.4.2.

ITQ 42

a) Inserting the appropriate values into the equation, to find the value of g_A, the surface gravity of Ariel (remembering to quote Ariel's radius, r, in metres), we get:

$$g_A = \frac{6.67 \times 10^{-11} \, \text{m}^3 \, \text{s}^{-2} \, \text{kg}^{-1} \times 13.5 \times 10^{20} \, \text{kg}}{(5.79 \times 10^5)^2 \, \text{m}^2}$$

$$= \frac{6.67 \times 13.5 \times 10^9 \, \text{m}^3 \, \text{s}^{-2}}{33.5 \times 10^{10} \, \text{m}^2} = 2.7 \times 10^{-1} \, \text{m s}^{-2}.$$

b) Inserting the values given for the ammonia–water eutectic melt in Table 5.7 and the value of Ariel's surface gravity that we have just calculated into Equation 5.6 in order to find the mobility index, I, of such a melt on Ariel, we get:

$$I = \frac{0.95 \times 10^3 \, \text{kg m}^{-3} \times 2.7 \times 10^{-1} \, \text{m s}^{-2}}{1 \times 10^2 \, \text{kg m}^{-1} \, \text{s}^{-1}}$$

$$= \frac{2.56 \times 10^2 \, \text{kg m}^{-2} \, \text{s}^{-2}}{1 \times 10^2 \, \text{kg m}^{-1} \, \text{s}^{-1}} = 2.6 \, \text{m}^{-1} \, \text{s}^{-1}$$

c) Inserting the values for a basaltic melt on Earth into the same equation, we get:

$$I = \frac{2.7 \times 10^3 \, \text{kg m}^{-3} \times 9.8 \, \text{m s}^{-2}}{2.7 \times 10^2 \, \text{kg m}^{-1} \, \text{s}^{-1}} = \frac{2.65 \times 10^2 \, \text{kg m}^{-2} \, \text{s}^{-2}}{2.7 \times 10^2 \, \text{kg m}^{-1} \, \text{s}^{-1}} = 98 \, \text{m}^{-1} \, \text{s}^{-1}$$

The values for andesite yield $I = 12 \, \text{m}^{-1} \, \text{s}^{-1}$, and for rhyolite $I = 0.098$ $\text{m}^{-1} \, \text{s}^{-1}$.

Table 5.9   Completed version of Table 5.7

Melt	Dynamic viscosity, η ($\text{kg m}^{-1} \, \text{s}^{-1}$)	Density, ρ (kg m^{-3})	Surface gravity, g (m s^{-2})	Mobility index, I ($\text{m}^{-1} \, \text{s}^{-1}$)
Ammonia–water eutectic on Ariel	1×10^2	0.95×10^3	0.27	2.6
Basalt on Earth	2.7×10^2	2.7×10^3	9.8	98
Andesite on Earth	2×10^3	2.5×10^3	9.8	12
Rhyolite on Earth	2.4×10^5	2.4×10^3	9.8	0.098

d) With a mobility index of $2.6 \, \text{m}^{-1} \, \text{s}^{-1}$, the ammonia–water eutectic melt on Ariel would be expected to show flow characteristics intermediate between andesite on Earth (mobility index $12 \, \text{m}^{-1} \, \text{s}^{-1}$) and rhyolite on Earth (mobility index $0.098 \, \text{m}^{-1} \, \text{s}^{-1}$). It would be about 5 times less 'mobile' than andesite on Earth, but about 25 times more 'mobile' than rhyolite on Earth.

SAQ ANSWERS AND COMMENTS

SAQ 1

The oldest oceanic crust is only a little older than 160 Ma (Plate 2.1, Block 2, Section 2.2.4), so whereas sea-floor magnetic stripes can be used to back-track the relative positions of continents at times since then, there are no comparable data available to track the motions of continents as they moved together to form Pangaea, which was complete by about 300 Ma ago.

SAQ 2

These regions of continental crust are along passive continental margins, which have been thinned and stretched by extensional (normal) faulting and have cooled (and therefore contracted) over time. Both thinning/stretching and cooling/contraction lead to isostatic subsidence, and hence flooding by shallow seas.

SAQ 3

Although this statement is consistent with one of the curves plotted on Figure 5.7, other curves suggest that continental crust may not have formed until after 4 000 Ma ago, and its growth rate may not have been steadily diminishing, but subject to episodic spurts in the Late Archaean and Early Proterozoic.

SAQ 4

(a) The internal heat generation in a chondritic mantle amounts to 4.06×10^{24} kg $\times 0.5 \times 10^{-11}$ W kg^{-1} = 2.03×10^{13} W. Together with the heat introduced from the core, the total heat input is $(2.03 \times 10^{13}) + (2 \times 10^{12})$ = 2.23×10^{13} W. This is less than the heat loss (3.6×10^{13} W), so a chondritic mantle would be cooling down.

(b) The net heat loss rate is 3.6×10^{13} W $- 2.23 \times 10^{13}$ W = 1.37×10^{13} W. Using Equation 5.5 and remembering that W = J s^{-1}:

$$1.37 \times 10^{13} \text{ J s}^{-1} = 4.06 \times 10^{24} \text{ kg} \times 1\,000 \text{ J kg}^{-1} \text{ K}^{-1} \times \text{ rate of}$$
$$\text{temperature change, } T'$$

$$T' = \frac{1.37 \times 10^{13} \text{ J s}^{-1}}{4.06 \times 10^{24} \text{ kg} \times 1\,000 \text{ J kg}^{-1} \text{ K}^{-1}}$$

$$= \frac{1.37 \times 10^{13} \text{ s}^{-1}}{4.06 \times 10^{27} \text{ K}^{-1}}$$

$$= 3.37 \times 10^{-15} \text{ K s}^{-1}, \text{ or } 106 \text{ K Ga}^{-1}.$$

(c) A chondritic mantle cools down slower than a peridotite mantle (~ 200 K Ga^{-1}) because chondrites have a greater heat generation.

SAQ 5

The basalts, which by definition have lower MgO contents than komatiite, may have been produced by partial melting in a cooler region of the mantle adjacent to that which produced the komatiites. A second possibility is that the basalts were produced by fractional crystallization of komatiite (cf. Block 3). Thirdly, the basalts could be the result of komatiite becoming contaminated by crustal rocks which have low melting temperatures and low MgO contents but high SiO_2 contents (Block 4). All three of these possible mechanisms find support in the chemical signatures of real basalt–komatiite suites.

SAQ 6

The mare basins were created during the late heavy bombardment, but were filled by lava subsequently. Thus, the craters on mare surfaces can be due only to the post-late heavy bombardment rain of meteorite, asteroid and cometary material.

SAQ 7

It is not possible, because the size–frequency distributions of craters at Jupiter and beyond are unlike those found in the inner Solar System. This can be taken to indicate that different populations of impactors must have been responsible for the cratering on the Moon and the satellite of Saturn referred to in the question.

(*Comment:* Within the inner Solar System *only*, we can use crater densities to deduce relative ages of surfaces on different bodies, and, taking advantage of the radiometric calibration of the lunar time-scale we can even suggest absolute ages. Elsewhere, crater statistics can be relied on only to compare relative ages of different surfaces on a *single* body, or (at best) between satellites of the same planet.)

SAQ 8

The highlands are densely covered by craters dating from the late heavy bombardment, and the major impact basins are superimposed on highland crust. This places the highlands firmly in the pre-Imbrian time period. Radiometric dates of 4420 Ma for the oldest highland anorthosites confirm this. (However, this is not to say that there is no ejecta of Imbrian and post-Imbrian times distributed across the highlands, and indeed several ejecta sheets can be mapped.)

SAQ 9

Its composition is similar to that of the other terrestrial planets (Mercury, Venus, Earth and Mars), as a result of forming from material that condensed in the same region of the solar nebula. The Moon and each of these planets is large enough (unlike any asteroid) to have experienced a long thermally active history. From the point of view of a geologist, geochemist or geophysicist, it is incidental that the Moon happens to be in orbit around another planet (the Earth). Except for Triton, the major satellites of the outer planets probably grew within a disc of gas and dust surrounding each planet, whereas the most likely explanation of the Moon's origin is that it was the result of a late giant impact by a planetary embryo onto the Earth (Block 1, Section 1.3.1).

SAQ 10

(a) Europa's density of 3.0×10^3 kg m^{-3} is the same as that of silicates used in the answer to ITQ 13. According to the formula used in the answer for ITQ 13, Europa's fraction of rock by volume (x) must be 1 so its fraction of ice ($1 - x$) is 0. However, the density of silicates within a large satellite such as Europa could slightly exceed 3.0×10^3 kg m^{-3} (through self-compression), which would make x slightly less than 1. Remember, though, that if the rock were hydrated its density would *decrease*! All we can really say is that Europa is unlikely to contain a significant fraction of ice.

(b) Figure 5.25 shows clearly that Europa's surface is dominated by ice. The density data show that this must a comparatively thin layer. (The ice could be no more than a surface skin, though in fact it is probably anything up to 100 km thick.)

(c) Io must have formed in the interior of the gas and dust cloud around the proto-Jupiter (i.e. in the proto-jovian nebula), where it was too hot for water to condense. Ganymede formed further out where the temperature was not significantly elevated by radiation from the proto-Jupiter, allowing a significant fraction of water to condense (as you can tell from its low density in Table 5.2). Europa orbits between these two, and presumably formed under conditions intermediate between those affecting Io and Ganymede; thus, a small amount of water was able to condense and accrete onto Europa without being lost.

SAQ 11

(a) In this case, the primary crust (see Section 5.3) is the lunar highlands, which are believed to have crystallized from a magma ocean.

(b) The Moon's secondary crust (see Section 5.3) is represented principally by the mare basalts, which formed by partial melting of the mantle.

SAQ 12

Below 1 000 km depth, S-waves are attenuated to the point of non-detectability, and P-wave velocity decreases. This suggests that an asthenosphere (which may be partially molten) begins at about this depth. Most deep moonquakes occur at 800–1 000 km, which by definition is within the seismogenic lithosphere (Block 1, Section 1.13); the lack of moonquakes at significantly greater depths suggests that 1 000 km is near the base of the lithosphere.

SAQ 13

Deep moonquakes are correlated with tides, and so they may tend to occur only within the Earth-facing hemisphere. They are also weak, and so ones more distant from the network of seismic stations would be harder to detect. Seismic waves from sources on the far side would have to travel through the asthenosphere (and, depending on location, through the core if it exists) where even P-waves are likely to be attenuated to some extent.

SAQ 14

(a) This is just a matter of simple manipulation of powers of ten. The Moon's annual seismic energy release divided by that of the Earth is 10^{11} J$/10^{18}$ J, which is $10^{11-18} = 10^{-7} = 1/10^7$ We hope that by this time in the Course you had no bother with this part of the question!

(b) The most basic reason is that the Moon's mass is about 80 times less than the Earth's, so, other things being equal, we would expect a reduction in annual seismic energy release by a similar amount. A scientifically more important factor is that the Moon has a much thicker lithosphere than the Earth, and this is not broken into tectonic plates. A corollary of this is that convection in the Moon's asthenosphere is unable to cause stresses adequate to fracture the lithosphere. Most moonquakes are a result of tidal stresses, rather than tectonic forces

SAQ 15

(a) The lunar highland crust is anorthositic in composition, and the most abundant mineral is plagioclase feldspar. The curves in Figure 5.32 show simply the preferential capture of Eu^{2+} by feldspar from the magma ocean, leaving the residual liquid (subsequently solidified before becoming the source from which the mare basalts were later derived by partial melting) with a corresponding negative Eu anomaly.

(b)   Block 3, Table 3.7, shows that the partition coefficient K_D between plagioclase and a basaltic melt is 0.07 for Rb and 2.2 for Sr. As discussed in Block 3, Section 3.7.1, this means that during fractional crystallization (as when the anorthositic lunar highland crust crystallized out of the magma ocean) Sr will be concentrated into the anorthite crystals ($K_D > 1$) and Rb into the melt ($K_D < 1$); thus, the lunar highlands should have a lower Rb/Sr than the mare basalt source region (as indeed they do).

SAQ 16

The enrichment in refractories and depletion in volatiles are part of the same phenomenon. If the material that now forms the Moon is the result of vaporization of a late-arriving planetary embryo and part of the mantle of the primitive Earth (Block 1, Figure 1.6), then volatiles will have been lost, and refractories retained. Upon impact, the core of the embryo is accreted onto the Earth's core, and the Moon is constructed from parts of the mantles of the two bodies. This mantle material will have been depleted in siderophile elements by the previous core-forming events, but this does not account for the difference between the mantles of the Earth and the Moon. To explain this requires a small iron-rich core to have formed within the Moon that would have caused a further phase of siderophile depletion in the Moon's mantle.

SAQ 17

(a)   On Earth, oceanic lithosphere (which is relatively dense) is usually subducted below continental lithosphere (which is less dense). The hypsographic plot for Venus can be interpreted to mean that Venus has no distinction between oceanic and continental crust. If this is correct, then there are no sharp crustal density differences on Venus, so that subduction would have to occur within crust of similar density, which would reduce the slab-pull force.

(b)   The high surface temperature on Venus and associated thermal expansion means that the buoyancy contrast between young (warmer) and old (cooler) lithosphere on Venus would be less than on the Earth, which would also inhibit subduction.

SAQ 18

The structure of Venus must be inferred without the benefit of seismic data such as we have for the Moon, and we are forced to rely on geochemical arguments and models based on density, moment of inertia and magnetic data.

SAQ 19

So far as we can tell, there is no plate tectonics on Venus, and if this is correct then plate recycling is the least important method of heat transfer on Venus (in fact, it may not operate at all). If we believe the situation portrayed in Figure 5.41, melts can only reach the surface in regions of enhanced heat flow, such as above mantle plumes. There appears to be good evidence for such hot spots in the coronae and other volcanic provinces on Venus, so clearly hot-spot volcanism is a contender. However, no active volcanoes on Venus have been observed, and the age of the surface (spanning 0–800 Ma according to ITQ 26, giving an average age several times older than the Earth's ocean floor), suggests that volcanism is not likely to be the most important mechanism. Most scientists are in fact agreed that Venus's heat transfer must be dominated by conduction through the lithosphere.

SAQ 20

(a)   Venus has degassed less than the Earth ever did (as shown by argon isotopes in the atmosphere); moreover, water in Venus's atmosphere would be broken down by ultraviolet radiation into hydrogen (which would escape to space) and oxygen.

(b)   According to the model presented in Figure 5.41, hydrated partial melts in the mantle would tend to crystallize as hornblende-bearing eclogite at a depth of around 100 km. This would prevent water being carried into the uppermost mantle, which therefore is drier (and, consequently, more rigid) than the Earth's, even in the convecting (asthenosphere) region.

SAQ 21

Mars has a lower density (even allowing for self-compression), which is compatible with oxidation of its iron content. The modelled composition of Mars based on analyses of SNC meteorites suggests that Mars's mantle plus crust is richer in oxidized iron than the Earth's. (*Note*: It would be wrong to suggest greater oxidation of material condensing from the solar nebula (as noted in the first paragraph of Section 5.9) as *evidence*. This is simply a *theory* that is consistent with the suggestion that Mars is more oxidized than the Earth.)

SAQ 22

(a)   The Hellas basin is the largest and one of the oldest of several low-lying basins on Mars that are probably the result of impacts, similar to the mare-forming impacts on the Moon. If the low-lying northern hemisphere represents an enormous impact basin, then the Hellas basin is only the second largest on Mars.

(b)   The Tharsis bulge is an updomed region of the northern hemisphere plains, upon which are constructed many of Mars's volcanoes, including the youngest, Olympus Mons. It may overlie the site of a persistent mantle plume.

SAQ 23

If it is the pressure of the overlying lithosphere that forces magma to the surface from a source just below the base of the lithosphere, then this pressure must be great enough to match the pressure at the base of a magma column extending from the surface to the base of the lithosphere. The calculation you did in ITQ 30 suggests that Mars's lithosphere in the region of Olympus Mons must be about 170 km thick. As this may lie over a mantle plume, which presumably heats the base of the lithosphere, the lithospheric thickness may be even greater elsewhere.

SAQ 24

(a)   The bulge must have been initiated very early in the history of the plains area within which it lies, because if any channel systems were older than the inception of the bulge we would expect the downstream slopes on some of them to have become reversed.

(b)   Most features that formed so long ago, like the Hellas basin, reached isostatic equilibrium because the lithosphere was still sufficiently thin and weak to allow this. However, the Tharsis bulge has been added to by volcanism up to the present day, so the bulge continued to grow *after* the lithosphere became unable to respond isostatically.

SAQ 25

(a)   There are no signs of any plate recycling, and volcanism, if it still continues, is restricted to Olympus Mons and possibly a few other locations. Most of Mars's heat must be lost by conduction through the lithosphere, so the rank listing in order of decreasing present-day importance as heat loss mechanisms is conduction, hot-spot volcanism (a negligible fraction), plate recycling (probably no contribution at all).

(b)   At 3 800 Ma, the rate of global heat loss was probably greater than today (because of greater radiogenic heat production and possibly heat inherited from the time of Mars's formation). Volcanism was more widespread at that time, but it is hard to say whether or not it was more important than conduction. There is no clear evidence that there ever was any plate recycling; the crustal dichotomy *could* represent the 'continental' part of a plate in the heavily cratered southern hemisphere and the 'oceanic' part of the same plate in the northern hemisphere, which would imply some sea-floor spreading, but there are no obvious fossilized subduction zones, and there are many alternative possibilities. The *most likely* rank listing in order of decreasing importance as heat loss mechanisms 3 800 Ma ago is the same as at present, though volcanism probably contributed a greater fraction to the total than now, with plate recycling probably a poor third.

SAQ 26

(a)   By dividing the total accretional energy available (7×10^{28} J) by the time (1.4×10^{17} s), you should arrive at a value of $5 \times 10^{11} \, \mathrm{J \, s^{-1}}$, which is $5 \times 10^{11} \, \mathrm{W}$.

(b)   This is a little over one-hundredth of Io's total power output at present.

(c)   Accretional heat should actually be lost fastest soon after accretion, and so the contribution of accretional heating to the total heat output of Io at present must be less than you have just calculated, and therefore be an even less significant heat source compared to tidal dissipation than the comparison in (b) suggests.

SAQ 27

The plumes are entirely compatible with lithospheric heat transport by rising magmas. Molten silicates were suggested as the immediate heat source for vaporizing sulphur in the higher temperature (Pele-type) plumes. The lower temperature plumes are more likely to be sulphur dioxide vaporized by the intrusion of molten sulphur; however, it would be reasonable to assume that this sulphur was initially melted by the intrusion of silicate magma into the sulphur source region.

SAQ 28

In view of all the volcanism on Io, its crust is clearly not primary (i.e. the result of crystallization from a magma ocean). It could be a secondary crust (formed by partial melting of the mantle), but as there has probably been a prolonged history of volcanism on Io its present crust is more likely to be recycled material. If the rate of resurfacing necessary to remove impact craters ($>1 \, \mathrm{mm \, yr^{-1}}$; Section 5.10.2) can be taken as representative of a long-term rate, then a crustal thickness of some 4 500 km would have built up over the lifetime of the Solar System. This is clearly impossible, so the base of the crust must be being removed (presumably by remelting) at a comparable rate to that at which the surface is being added to. In other words, it seems that the present crust *must* be recycled material, and is therefore tertiary. (Primary, secondary and tertiary crust were defined in Section 5.3, pp. 13–14.)

SAQ 29

You should have ended up with something looking like Figure 5.75. The preponderance of hot-spot volcanism as the main mechanism of heat loss on Io was established in Section 5.10.5, so the point for Io should lie close to the hot-spot volcanism apex of this triangle, and (in the absence of plate tectonics) exactly on the edge opposite to the plate recycling apex (to indicate that plate recycling probably plays no role at all). Do not worry if your other points are not in exactly the same positions, so long as they are in the right locations relative to each other; for example, Mars loses a greater proportion of heat by hot-spot volcanism than the Moon, but it is not clear how much greater, since the extent of present-day activity of Olympus Mons (averaged out over the current 100 million years or so) is unknown (see the answer to SAQ 25). The least certain point on this plot is the one for Venus (see the answer to SAQ 19), despite (or perhaps because of!) all the recent data from Magellan; if the average age of Venus's surface has been correctly worked out (Section 5.8.3), then volcanoes cannot account for a large proportion of the heat loss, and nor does much plate recycling seem to occur, which leaves lithospheric conduction as the most important mechanism. However, if the Venus data have been wrongly interpreted, then it is quite possible that Venus ought to be plotted rather further from the lithospheric conduction apex.

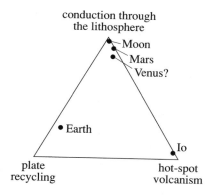

Figure 5.75 Completed version of Figure 5.60, for the answer to SAQ 29.

SAQ 30

The temperature in the solar nebula was too high at the radius of Jupiter's orbit for ammonia to condense. There was probably also an additional temperature increase due to Europa's proximity to Jupiter (see SAQ 10(c)), which would make the condensation of ammonia even more unlikely.

SAQ 31

Because of its smaller radius and lower density, it should be obvious that the surface gravity on Miranda is less than that on Ariel; thus, a melt of the same dynamic viscosity and density would be less mobile on Miranda than on Ariel. If you insert the values of mass and radius given in this Table into Equation 1.34, you can calculate that the surface gravity on Miranda, and hence (because, for the same dynamic viscosity and density, they are proportional) the mobility index of the ammonia–water eutectic melt on Miranda, are each about 35% that on Ariel.

SAQ 32

(a) Your answer should look something like that in Table 5.10. You should refer back to the text on each specific body if you got it badly wrong. The completed Table is not meant to imply that chemical data for the Moon are as good as those for the Earth, merely that they are much better than for Mars and Venus. Note that the discussion of the *Voyager* data density on both Io and the icy satellites appears in Section 5.6.

Table 5.10 Completed version of Table 5.8, for SAQ 32.

	Chemical sampling	Seismic measurements	Moment of inertia measurements	Density determinations
Earth	✓	✓	✓	✓
Moon	✓	(✓)	✓	✓
Venus	(✓)		(✓)	✓
Mars	(✓)		(✓)	✓
Io				(✓)
Icy satellites				(✓)

(b) Reflectance spectra, which demonstrate the presence of H_2O ice on icy satellites (see Figure 5.25), with the addition of nitrogen in the case of Triton (Section 5.11.4). The spectrum of Io contains evidence of both sulphur and sulphur dioxide (Section 5.10.2). (*Note*: Although we have not discussed it in this Course, reflectance spectra can also give clues to the mineralogy of silicate surfaces.)

SUGGESTIONS FOR FURTHER READING

A book containing chapters written by specialist authors covering all aspects of the Solar System (not just the planetary bodies considered in this Block), but at a superficial level, is: Beatty, J. K. and Chaikin, A. (eds) (1990) *The New Solar System* (3rd edn), Sky Publishing Corporation & Cambridge University Press.

Another general book is: Greeley, R. (1987) *Planetary Landscapes* (revised edition), Allen & Unwin. This concentrates on surface processes, and surface manifestations of internal processes, on the terrestrial planets.

There are several books available that treat the Moon in some detail. Two that extend the discussion beyond the point reached in this Course are: Cadogan, C. H. (1981) *The Moon — Our Sister Planet*, Cambridge University Press, and Taylor, S. R. (1982) *Planetary Science: A Lunar Perspective*, Lunar and Planetary Institute. The story of how our knowledge and understanding of the Moon developed is set alongside the history of lunar exploration in a highly readable non-technical account in: Wilhelms, D. E. (1993) *To a Rocky Moon: A Geologist's History of Lunar Exploration*, The University of Arizona Press.

The definitive students' book on Venus has yet to be written, but for Mars the following is recommended: Cattermole, P. (1992) *Mars; The Story of the Red Planet*, Chapman & Hall. An excellent review of Mars geology, but not much on its interior.

Finally, if you enjoyed Sections 5.6, 5.10 and 5.11, you should read the following, which discusses the geology and evolution of Io and all the large (>200 km radius) icy satellites: Rothery, D. A. (1992) *Satellites of the Outer Planets: Worlds in Their Own Right*, Oxford University Press.

ACKNOWLEDGEMENTS

I wish to thank the following for their helpful, insightful and stimulating comments and discussion of various parts of this Block during drafting: Richard Thorpe and Geoff Brown (both now sadly deceased), David Darbishire, Peter Francis, Colin Hayes, Simon Kelley, Ray Macdonald, Harry Pinkerton, Val Russell and the S267 Course Team. I am also grateful to Dick Carlton for assistance in enhancing some of the spacecraft images.

Grateful acknowledgement is also made to the following sources for permission to reproduce material in this text:

Figures 5.1 and 5.4 R. D. Nance, T. R. Worsley and J. B. Moody (1988) 'The supercontinental cycle', *Scientific American*, July, copyright © 1988 W. H. Freeman & Co.; *Figure 5.5a,b* B. F. Windley (1977) *The Evolving Continents*, Wiley, reprinted by permission of the publishers; *Figure 5.5c* C. J. Talbot (1973) 'A plate tectonic model for the Archaean crust', *Phil. Trans. Roy. Soc. Lond.*, **A273**, The Royal Society; *Figure 5.6* L. D. Ashwal (1989) 'Introduction to growth of the continental crust', *Tectonophysics Special Issue*, **161**, Elsevier; *Figure 5.7* M. S. Drummond and M. J. Defant (1990) *Journal of Geophysical Research*, **95**, American Geophysical Union; *Figure 5.10* M. H. Jackson and H. N. Pollack (1984) *Journal of Geophysical Research*, **89**, American Geophysical Union; *Figure 5.12* G. H. Miller *et al.* (1991) *Journal of Geophysical Research*, **96**, American Geophysical Union; *Figure 5.13* F. J. Doyle, US National Space Science Data Center; *Figure 5.16* Johnson Space Center Mapping Sciences Laboratory; *Figure 5.17* D. E. Wilhelms and D. E. Davis (1971) 'Two former faces of the Moon', *Icarus*, **15**, No. 3, Dec. © 1972 Academic Press; *Figure 5.21* Canada Centre for Remote Sensing, Department of Energy, Mines & Resources, Ottawa; *Figures 5.22 and 5.33* B. C. Murray, US National Space Science Data Center; *Figures 5.23, 5.38, 5.54, 5.62, 5.63, 5.66 and 5.68* B. A. Smith, US National Space Science Data Center; *Figure 5.27* M. N. Toksöz *et al.* (1974) 'Structure of the Moon', *Review of Geophysics and Space Physics*, **12**, No. 1, American Geophysical Union; *Figures 5.31 and 5.44* NASA; *Figure 5.35* NASA Planetary Data Center; *Figures 5.36–40* G. H. Pettengill, US National Space Science Data Center; *Figures 5.45, 5.48 and 5.49* M. H. Carr, US National Space Science Data Center; *Figure 5.55* J. R. Spencer; *Figure 5.65* US Geological Survey.

INDEX FOR BLOCK 5

(**bold** entries are to key terms; *italic* entries are to tables and figures)

Plate 5.1 A close-up view of Neptune's largest satellite, Triton. Its surface
is too young to have been greatly scarred by impacts, although there are a
few impact craters visible in the lower part of this picture. The surface
appears to be a frozen mixture of water-ice and methane, sculpted by icy
volcanic processes of the sort you will learn about in Section 5.11, and is
overlain by a bright polar cap of frozen nitrogen. (*US Geological Survey,
Flagstaff*)

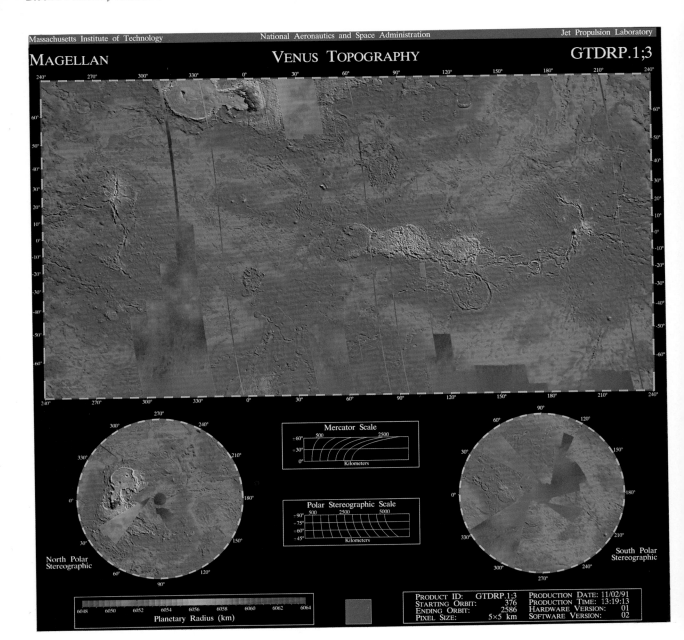

Plate 5.2   The topography of Venus as mapped by the *Magellan* radar altimeter. Blue colours have been assigned to the lowest elevations, green to intermediate elevations and yellows and reds to the highest elevations, relative to the global datum. The grey strips cover areas where data were lacking when this compilation was made. The latitude and longitude co-ordinates can be used to locate the images shown in several of the accompanying Plates and the Figures. There are two major highland regions: Ishtar Terra, high in the northern hemisphere, and Aphrodite Terra, in the equatorial part of the eastern hemisphere. (*NASA/Jet Propulsion Laboratory*)

Plate 5.3 A global view of Venus (centred on the equator at 180° longitude) produced from *Magellan* radar images. Colour has been added to the radar image mosaic to match the surface colour recorded by the *Venera* landers. The highlands of Aphrodite Terra (see Plate 5.2) correspond with the east–west radar-bright belt just to the south of the equator. (*Bradford A. Smith, US National Space Science Data Center*)

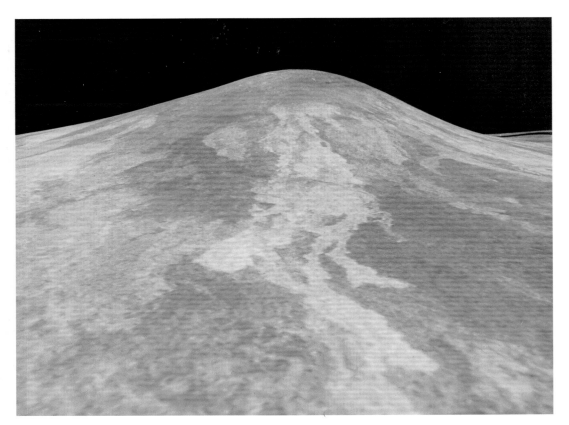

Plate 5.4   An artificially constructed and vertically exaggerated perspective view of the volcano Sif Mons on Venus, made by combining a radar image from *Magellan* with topographic data. Bright lava flows are clearly seen, heading directly towards the viewpoint. Colour has been added to match the surface colour recorded by the *Venera* landers. This view is about 200 km wide and shows a height range of about 2 km. You should be able to identify Sif Mons on Plate 5.2, near 20° N and 0° longitude. The gentle slopes of this volcano are strongly suggestive of low-viscosity lava such as basalt. (*Bradford A. Smith, US National Space Science Data Center*)

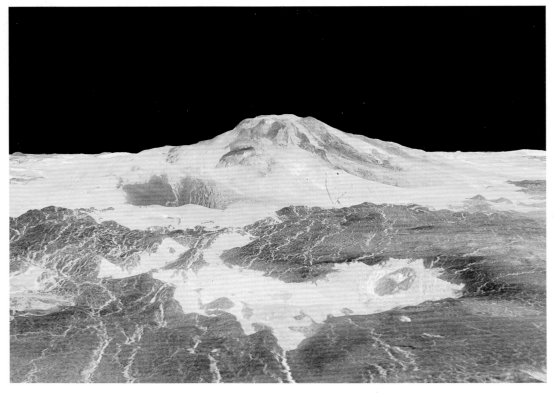

Plate 5.5   A perspective view of the 5 km high volcano Maat Mons (0° N, 195° E) on Venus, constructed as for Plate 5.4. The impact crater in the right foreground is 23 km in diameter. (*Bradford A. Smith, US National Space Science Data Center*)

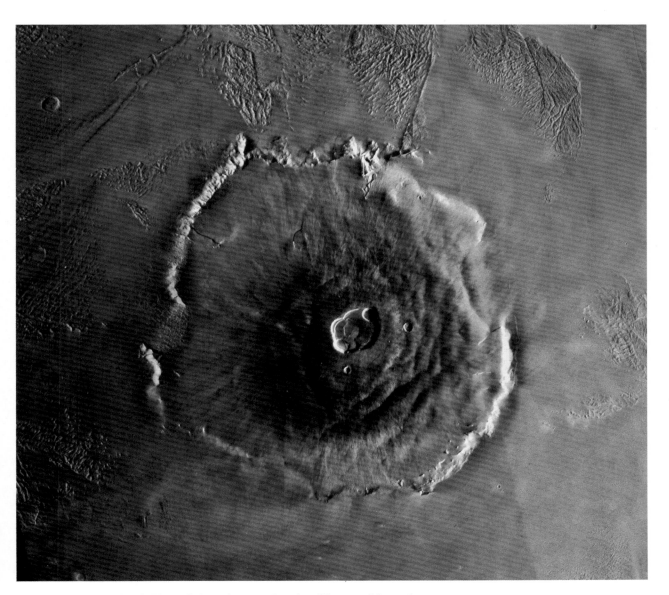

Plate 5.6   A mosaic of *Viking Orbiter* images showing Olympus Mons, the highest and youngest volcano on Mars. The area shown is *c.* 1 000 km. across (*US Geological Survey, Flagstaff*)

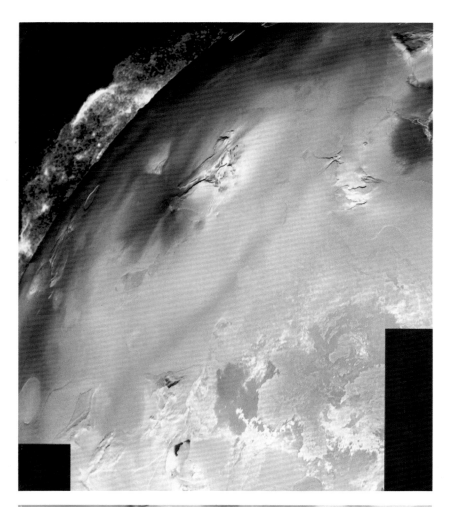

Plate 5.7   An oblique view across part of Io showing the active volcano Pele. The plume reaches about 300 km above the volcano. It is clearly visible against the blackness of space, above the horizon, because its appearance has been enhanced to make it show up. (*US Geological Survey, Flagstaff*)

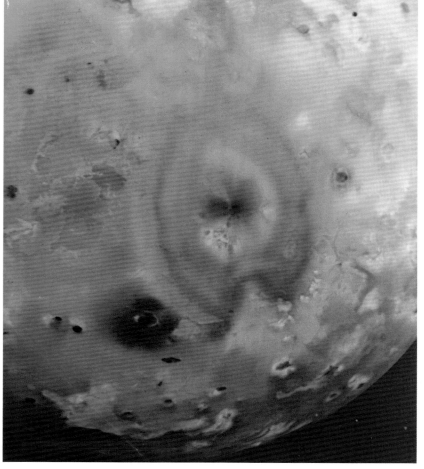

Plate 5.8   A near-vertical view downwards through the eruption plume from Pele, on Io. Fall-out from the plume is forming the dark oval-shaped ring deposit on the surface. This view is about 1 850 km across. (*NASA*)